YEAR 2000

The Inside Story of Y2K Panic and the Greatest Cooperative Effort Ever

Nancy P. James

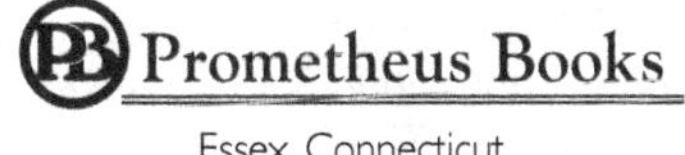

Essex, Connecticut

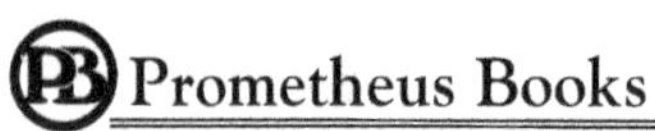

An imprint of The Globe Pequot Publishing Group, Inc.
64 South Main Street
Essex, CT 06426
www.globepequot.com

Distributed by NATIONAL BOOK NETWORK

British Library Cataloguing in Publication Information Available

Library of Congress Cataloging-in-Publication Data Available

ISBN 9781493085613 (pbk. : alk. paper) | ISBN 9781493085620 (epub)

∞™ The paper used in this publication meets the minimum requirements of American National Standard for Information Sciences—Permanence of Paper for Printed Library Materials, ANSI/NISO Z39.48-1992

CONTENTS

INTRODUCTION: "Y2K"

The Year 2000 millennium computer technology crisis, known as Y2K, was a distinctive time in modern culture. The problem: What would happen when the use of computerized two-digit date fields used through all of the latter half of the 1900s encountered the date 01/01/2000?

No one knew.

Worst-case scenarios predicted a complete collapse of our nation's—perhaps our planet's—infrastructure, causing massive deaths, casualties, and destruction in mid-winter, the worst possible time for a crisis in the northern hemisphere. Year 2000, like COVID-19, was a completely new problem requiring enormous resources for discovery and remediation. Everything was an unknown—an estimated $600 billion to $1 trillion problem. Infrastructure critical to civilization—including heat, electricity, water, and sanitation—were at risk, all complete unknowns. Those unknowns included whether the date issue would actually cause a problem; how to identify affected dates in billions of lines of code; how to most economically upgrade and test computer code before the turn of the millennium; and whether it would all work on January 1, 2000.

The human cost of failure was given the majority of worst-case media attention, but during the final decade before the new millennium, 1990 to 1999, other competing pressures foresaw the possibility of a complete collapse of our economic structure. The U.S. tort bar (law firms who specialize in filing civil suits for damages) anticipated a trillion dollars in Y2K litigation damages and expenses, with insurance proceeds at the head of their funding list. Total U.S. casualty/property insurance reserves at the time were just under $400 billion. Imagine the complete collapse of the global insurance market—no home, automobile, hospital, manufacturing, workers' compensation, marine, aviation, medical, or professional malpractice insurance. Global economies would come to a standstill. Chaos and anarchy would surely reign. However, no media attention was given to the possible collapse of the insurance sector and consequences to the global economy. Pausing to note such an oversight is one reason why Y2K is an interesting story to revisit.

There was fear of an accidental nuclear arms deployment. There was fear of monetary systems being jeopardized, as well as the fear of infrastructure collapse. There were simultaneously many comic moments as well; voluble hysterics, strident survivalists, unstudied opinions, and nonsensical, arcane solutions. Y2K "survival guides" were filled with suggestions to stockpile water and food, have first aid kits ready, and more to prepare for the possible end of the world once the clock struck midnight.

Yet the world didn't end on January 1, 2000. In fact, most people rang in the new year with the perception that nothing happened at all.

Uniquely, the global Y2K struggle did not appear to have the typical tragic side with great numbers of lives being lost in the effort. This positive outcome was not a stroke of luck, nor was it because people overestimated or exaggerated Y2K risk. It was only possible because people across industries and sectors worked to offset disaster. For all of the rivalries and self-interest that is documented in this book, the fundamental successful outcome was sought and worked on faithfully by each sector.

When you look at this story from the lens of the people who did not know the outcome, you see the depth and the influence of their voices. The noble and the lesser.

Y2K was not a heroic tale of brilliant intervention at the eleventh hour; no mounted cavalry. From the public perspective, it seemed quite dull, possibly nonexistent to many. In fact it was a tale of hundreds of thousands of hours of tedious work, skill, and determination. Y2K is a story of human nature, of the activists, the optimists and pessimists, the actors and spectators. I was one of those actors and this is my story.

The concept of 20/20 hindsight has its purpose: Not to judge—well, maybe to judge—but more importantly to hold the actors involved accountable and to offer perspective on an event that was, by all accounts, ultimately successful.

WHAT *WAS* Y2K?

"The Year 2000 Computer Problem," "The Millennium Bug," "Y2K Bug," "The Millennium Time Bomb" "Time Bomb 2000," "The Year 2000 Problem," "Armageddon," "The Fourth Horseman of the Apocalypse," "TEOTWAWKI" (The End of the World as We Know It, pronounced "tee-OH-tawa-kee"), "El Niño of the Digital Age," and "Black Saturday" were among the ominous names for what proved to be a unique example of a distinct, single-issue problem with global ramifications.

In the early days of computers, software programmers coded almost all data to save expensive storage space. Clerical data-entry employees learned complicated coding structures to enter onto IBM 80-column cards or paper tape reels. Data, with the general exception of names and dates, used short alphanumeric characters to represent complex business information. Naturally, it was logical for such methods to use only two digits for year dates.

That meant that when the Year 2000 began, its "00" year indication would be mistaken for 1900 instead of 2000 when the new millennium arrived. The issue of six digits for month, day, and year (e.g., 010177 for January 1, 1977) would run amok when the calendar rolled over to 2000 (010100), confusing dates with the previous century, suggesting newborns were 100 years old! All of this was compounded by the fact that 2000 was a leap year; while years divisible by 100 are not leap years, years divisible by 400 are. Thus, the Year 2000 would have the extra day in February.

And so, "Y2K" was born.

Legacy computer systems for large, long-standing, venerable institutions had to be identified, analyzed, and remediated in some way, either by replacement or tedious, arcane patches. Financial institutions were among the most vulnerable with a multiplicity of regulators and overseers.

The very fact that this software date issue became universally known as the "Y2K Bug" acknowledged a built-in assumption that it was a flaw, a failure, most likely a failure of foresight. But it is important to pause and be reminded that reports of cost for one megabyte of data in 1970 was $3 million; in 1980, $64,000; in 1998, the cost was $5.[1] The "Bug," therefore, was not an oversight, nor shortsighted, or even a "bug" at all. In fact, it was a sound economic decision for its time.

That being said, bug or not, Y2K couldn't go unaddressed. The situation demanded widespread intervention before the clock changed to "00."

A TREMENDOUS STORY BEHIND A "NON-EVENT"

For over twenty years I have considered the success of Y2K millennium computer remediation as the greatest cooperative effort ever undertaken by humanity on the face of the earth. Cooperative, because every nation shared the same goal of keeping infrastructures operational—basic services for heat, electricity, municipal water and sewer, law enforcement, and communications. But the same old rivalries were very much present, playing out in struggles, blame, and strategies.

As my odyssey through the late 1990s began and continued, I amassed and saved thousands of documents about the coming millennium and the computer issues facing both the United States and the world. Going through four storage boxes to rediscover the range of thought, genius, and folly involving but a single, quantifiable issue offers insight into how the human mind and spirit work toward progress. In this case, ultimate cooperative success. Every player from coders to doctors, U.S. Congress to municipal officials, finance to the legal profession, had essentially the same goal, a successful outcome. Everyone had a pony in this race.

Of particular note during the years while I was writing this book, the absence of cyber warfare and cyber terrorists was specifically stark. Rogue nation-states were not watching for a strike opportunity nor sabotaging adversaries' Year 2000 remediation efforts with sophisticated cyber offensives, even while some were planning traditional New Year's coincident attacks. Early on New Year's Day, 2000, there was a very brief television report of a Y2K outage in the Pentagon's reconnaissance satellites that caused the loss of all data, including all spy satellite data. Not until January 4 was it reported that the outage had lasted for three days beginning 7:00 p.m. on December 31, 1999 (midnight GMT, January 1).[2] This was very troubling, considering that the United States reported sharing early-warning satellite data pre-millennium with both Russia and China to prevent uncertainty and panic over the possibility of an unknown attack if Y2K darkened those countries or ours.[3] The public will likely never know how tense our highest defense and military officials were regarding that Y2K satellite data blackout. Twenty-plus years ago, had our government and military leaders reacted differently, or had more sophisticated and far-reaching cyber warfare techniques been more widely available, that issue might have had catastrophic consequences for the United States.

Beyond our military, singularly unusual to this cooperative character of Y2K was the inclusion of every enterprise by virtue of its own house requiring Year 2000 re-engineering. Government entities, while enacting legislation to hold themselves harmless, were intensely laboring on internal Y2K compliance efforts. Attorneys were looking to the litigation potential while their back-office technology teams were laboring for smooth Year 2000 compliance. Insurers looking to avoid massive Y2K claims were anxious not to trip on their own internal Y2K remediation defenses. Those seeking to gain the most, Y2K consulting firms in this author's opinion, were uniquely facing the very same in-house issues on which they were laboring externally. Perhaps, indeed, the millennium offered the rare opportunity for sympathetic perspective from every sector.

Y2K AND ME

In the mid-1990s I began reading about Year 2000, later writing about it, and then became a part of it.

My work with the issues involved in the date change to January 1, 2000, started in the early 1990s when technology publications began to carry a growing number of articles about computer programming problems interpreting the Year 2000. As owner of a casualty/property insurance agency specializing in technology risks, my professional life was devoted to risk identification, analysis, and management. This is a chronicle of my experience during those pre-millennium years tackling the various aspects of risk analysis of those technology issues. It is also a look on the Great Recession ten years later when I attributed much of that recession to ongoing computer bugs from hastily upgraded replacements for Y2K compliance. Year 2000 caused corporate management to be particularly cautious when investing in new technologies following the stresses of the turn of the millennium. The issues of remediation for Year 2000 had long-lasting effects.

I began to be noticed on the Y2K liability issue in 1996, when I coauthored an article with attorney Barbara O'Donnell published in the *Boston Bar Journal*.[4] Around the same time, the Gartner Group would testify before Congress and widely publish estimated costs of a $600 billion U.S. Y2K remediation project, which preceded the American Bar Association's widely quoted $1 trillion estimated projection in litigation costs. Cost estimates of both compliance and litigation varied wildly with huge deltas: The Gartner Group was quoted in 1998 estimating global litigation costs from $1 trillion to $3.3 trillion.[5] The American Bar Association also quoted total repair and litigation costs as high as $3.5 trillion.[6]

My position on risks involved with Y2K software compliance was firmly that insurance proceeds would not respond to claims of remediation failures. As an insurance professional, my specific concerns turned to the tort bar's assertion that insurance policy coverage would likely respond to the $600 billion remediation costs and $1 trillion software failure litigation expenses and judgements. Such numbers would guarantee catastrophic consequences to the entire domestic casualty/property insurance market.

Merrill Lynch's Jay Cohen noted the same concern in a mid-June 1996, research report.[7] My concern over the tort bar's focus on asserting insurance proceeds to cover Y2K remediation expenses and litigation led to my mid-1998 letter to President Clinton's top Y2K official, John Koskinen, recommending a mediation alternative. "Can we establish a Millennium Mediation Council before the 'loss' to not only balance responsibilities, fees, and liabilities, but also to direct resources away from adversarial pursuits toward a solution? If we can, we can lead the globe on this matter," I wrote. My letter prompted a cordial telephone call from Mr. Koskinen to discuss my suggestion as well as a reply letter from him.* As my work and voice became more prominent, media began to take heed. The Associated Press quoted me on my

* See the appendix for a copy of my letter and Koskinen's reply.

position too: "[I]nsurance was never intended to give Year 2000 coverage. No premiums were ever collected for it."[8]

Dramatic variations in cost estimates for both domestic and global Y2K software remediation complicated the issue. There were so many unknowns about the scope of work, from lines of code, to viable solutions, to time frames. The Associated Press estimated a huge variable between $600 billion and $4.6 trillion for worldwide conversion costs.[9] Cap Gemini, a consulting firm, estimated that U.S. corporations would spend $520 billion on Y2K remediation.[10] While the $600 billion costs were more commonly used, Cap Gemini revised their estimates in 1999 up to $750 billion to $850 billion.[11] Insurance journal *National Underwriter* cited *Newsweek* estimating total global costs for remediation and litigation to reach as high as $1 trillion. The same article noted that Software Productivity Research in Boston put global remediation estimates at $3.6 trillion.[12] The London-based *Economist* also quoted SPR's estimate for their readers.[13] A. M. Best was noted as saying that next to catastrophic risk, Year 2000 would be the second most important rating issue for 1999 and forward.[14] In 1996 and earlier, the $600 billion Year 2000 remediation costs were thought to be global, later revised to estimate just the U.S. costs.[15] Those cost estimates were in comparison to the U.S. gross domestic product (GDP) of $7.3 trillion in 1998.[16] With estimates of the financial magnitude of Year 2000 compliance issues so disparate, there was reason to understand the nation's and the world's anxiety.

In our relatively early 1997 article for attorneys regarding Y2K computer liabilities, Attorney Barbara O'Donnell and I reviewed the details expected to cause computer malfunctions when the clocks turned over from midnight, December 31, 1999, to January 1, 2000. We provided estimates for government and business remediation expenses along with expert opinions as to the accuracy of those estimates. The role of both legal and insurance professionals for protection of clients followed for contracts, warranties, first- and third-party liabilities, and property and business interruption. Insurance response to various potential presumed causes of loss, possibly described as "natural disasters" or "failure to perform," as well as the application of other exposures to Y2K failures were outlined. We offered that Y2K losses, like previous high-expense national issues, could be reinterpreted by the courts to force insurers to pay claims for damages not contemplated by insurance policies.

We raised all of these issues before the American bar published an estimated $1 trillion in Y2K litigation costs, and before Y2K exclusions were drafted for attachment to insurance policies. I maintained my position that insurance proceeds would not be responding to Year 2000 compliance costs all through the millennium change. My rationale was based on several factors: Y2K remediation coverage expenses were clearly outside of several policy terms; timely notice requirements could not be met; rating and premium was never collected for the exposure. However, I had nagging fears that either Congress, or possibly the courts, would intervene to reinterpret Year 2000 software expenses in some novel terms that would clearly be a "covered cause of loss" in standard insurance policies. Should some court declare Y2K issues a "virus," for example, many insurance policies would be obligated to provide coverage for remediation and compliance costs, with that precedent opening billion-dollar flood gates. Insurance proceeds have long been thought of as a free pool of money just waiting for opportunity.

My cautions at the time were quoted in May 1998 by *Dow Jones News Service,* where I provocatively posed: "What if the government determines Y2K is a (computer) virus? Or, what if the government declares the Year 2000 fallout a natural disaster. . . . That could trigger coverage we otherwise hadn't anticipated."[17] My opinion has endured over the years, as I have watched courts assault contract terms to provide funding for much-needed costs associated with pollution cleanup, for example. Congressional renaming of natural disasters is, in my opinion, the fastest way for quasi-public (is this not insurance, after all?) non-tax-dollar funding for restoration. It's odd that our Washington representatives have not used it more frequently.

Our *Boston Bar Journal* article created considerable interest; as a consequence, I was asked to speak on the Technology Panel of the American Bar Association's TIPS Annual Meeting held in San Francisco in July 1997. Craig Stewart of the Boston law firm Palmer & Dodge, LLP, then serving on the 1997 TIPS Annual Meeting Program committee, had taken an interest in O'Donnell's and my article and asked if I might be willing to serve on the panel following a conference call where I offered my opinion that software litigation was relatively old news in comparison with the growing risks of both cyber-liability and Year 2000 issues.

Two fellow members of the panel, Daniel J. Langin, Esq., Media/Professional Insurance (Kansas City, Missouri), and Tom Cornwell, vice president, Chubb Insurance Company (Warren, New Jersey) were very important figures in my professional life. Dan Langin had been the attorney who assisted me in creating the first internet liability policy in the country a few months earlier, orchestrating the "war game" that Dan, his underwriters, and I arranged at my client's site in April 1997.[18] Chubb's Tom Cornwell was the first person to alert me to the cost considerations of data storage, which I have referenced frequently in my story. Tom noted that the measurement of Y2K remediation and compliance expenses would inevitably be far less than comparable data storage costs of four-digit date fields in legacy COBOL systems. I considered that fact, then and now, one of the most important factors for the coming Year 2000 tumult and fear, at least from an insurer's perspective.

Government, corporate, and media attention to the Year 2000 problem was growing as 1997 ended, with service firms alerting clients to Y2K. The Boston law firm of Palmer & Dodge published a "Client Alert" in November 1997, describing compliance effects on portfolio companies, investors, banks, borrowers, and lenders. Cost estimates were $.50 to $1.50 for modification of every line of code with a note that a "major insurance company's programs being reported to contain over 125 million lines of code."[19] (Later reports of coding cost estimates were: $2.95 H1-1999, $3.65 H2-1999, and $4.00 H1-2000.)

Barron's cited J. P. Morgan forecasting that 50 percent of information technology (IT) budgets would be sapped for Y2K until 2000. A Society of Information Management survey found that 75 percent of IT professionals believed Y2K costs would come from their regular IT budgets, warning that "something else will be sacrificed, and there will be consequences."[20] Those consequences surely were primarily security-related, as hackers were travelling full steam ahead while millennium remediation efforts distracted defense efforts.

HOW MEDIA FRAMING IMPACTED PUBLIC PERCEPTION

"Y2K is a crisis without precedent in human history," declared a mid-1998 feature article in *Byte Magazine*.[21] That provocative statement sets the stage for my story. At the end of the last century there was a singular moment in time, turning the clocks to January 1, 2000. From it emerged a worldwide project with a specific goal and an immovable fixed completion date: 01/01/2000.

Watching the Y2K software issue become a global crisis appeared to me to have more external influences than any era or area in history, with the exception of world wars. Those external influences encompassed every major sector of our contemporary lives: governments (federal, state, and local), courts, all finance and insurance sectors, the American bar, the internet, virtually every commercial enterprise, and our fundamental infrastructure that we took then, and still do, completely for granted. I wrote about those reflections and observations as it became clear that the otherwise staid insurance sector would be subject to more disruption than any other period in time. It was not a long game; the end was approaching.

Preceding my work by nearly three years was one of the early Year 2000 gurus, Peter de Jager, among the most prominent for his essay "Doomsday 2000" for *Computerworld* in 1993. De Jager was regarded by the *New York Times* "as the information-age equivalent of the midnight ride of Paul Revere." The *Times* quoted de Jager: "We and our computers were supposed to make life easier; this was our promise." However, he went on, "What we have delivered is a catastrophe."[22]

Adding to that, original equipment manufacturer (OEM) products containing software and electronic parts from a variety of sources further complicated matters for Y2K failure issues. OEM products included those upstream software and parts issues before Year 2000 compliance, with an added Y2K high-value layer of potential failure. However, the basic issues were related more to licenses and patents as opposed to quality control and product failure. Y2K just added another issue to handle.

Frustratingly, most lengthy reports continued to depict Y2K as primarily a programming error. Even while referencing the early mechanics and costs of data storage, nevertheless the issue was considered to be boiled down to programming "shortsightedness" as programmers of the 1960s onward were being faulted.

Respected media sources from all industries were in agreement on that matter, with an early article on Y2K in the June 1996 issue of *Service News* advocating technical and business group cooperation, which opened with "Mainframe programmers, not foreseeing the long shelf like of their business applications, cut corners by identifying years in two digits within their programs." With A-list firms of Keane, Gartner Group, Dunn & Bradstreet, and Alydaar Software featured, Kevin Schick, research director at Gartner Group, referenced 20 percent waste with redundant and never-used applications, saying "Not taking advantage of this opportunity to clean house is 'sinful.'"[23]

It was comforting to learn early on that theologians had taken a stand on corporate matters, at least regarding the millennium, in case they never did again. In a 1999 feature article in the *New*

York Times, the two-digit date field software coding was described as an "ethical lapse." The article went on to compare this programming "ethical lapse" to the regrets of nuclear physicists in developing the atomic bomb. Paul Strassman, a former Pentagon technology officer, predicted that chief information officers would one day be legally accountable for computing operations.[24]

But, the staggering cost differential of a megabyte of memory was rarely mentioned by the media. The Y2K software programming issue was declared by fiat, by many names, as nothing more than a blunder of varying magnitudes.

An early 1999 *Times* article reported Y2K issues with the Internal Revenue Service, referring to it as "the pesky software flaw," supporting the almost universal notion that the two-digit year date was an oversight or accident. "Charles O. Rossotti, the Commissioner of Internal Revenue, said 'Our task is like changing the engines in the plane while you are still flying.'" The IRS, with an estimated 100 million lines of code, at the end of the twentieth century was reported to "still be using huge reels of magnetic tape."[25]

Technology magazines in the mid- and late 1990s were filled with Year 2000 advice and admonitions. One article by a computer consultant Leland Freeman in *Software Magazine* likened Y2K to war scenarios, calling it a "Year 2000 Pearl Harbor." Using the combat analogy, Freeman likened Y2K more to a cold war than a world war, suggesting the 1950s strategy of "duck and cover" was the prevailing strategy to handle this crisis. National leadership coupled with a slow-to-act Congress suggested the need to declare "World War Y2K" and defuse the "Y2K bomb."[26] Looking back on the contemporaneous sequence of events, this early depiction of Year 2000 in terms of "Year 2000 Pearl Harbor," "doomsday," "catastrophe," and "pesky software flaw" surely fueled much of the panic and pressures that enveloped the issue from the outset.

Looking back, I have concluded that a single overwhelmingly common opinion caused much of the anxiety and consequential blame. Our culture consistently looks back with 20/20 hindsight;

it's what we do. But to program with two-digit year fields was good judgment at the time; we knew nothing different. It is important to note that, even while the silicon chip was invented as far back as 1961, most financial institutions systems were not on the cutting edge of technology. Because of the very nature of our vast financial sector networks and the necessity of reporting to regulators, financial sector computer systems were and continue to be slow to change.

Succinctly and skillfully summarized by Peter Worrall in 1997:

> More information has been produced in the last 30 years than in the past 5,000. Through innovation and adaptability, technology has jumped every hurdle put before it: in the 1950s valves were replaced by transistors; in the 1960s transistors were replaced by integrated circuits; and in the 1970s the microchip arrived. The 1980s saw an enormous increase in the power of the chip, combined with the launch of dedicated software programs and new operating systems. This brought about ultimate accessibility—the personal computer. As a result of these improvements, the price of unit storage today is down a staggering 99.9 percent compared to the 1970s. This means a typical organization using a two-digit format has saved $1 million per billion bytes of storage between 1963 and 1992 (in 1992 dollars).[27]

Year 2000 expert Capers Jones offered this metric in his 1998 abstract, "Dangerous Dates for Software Applications,"[28] where he described the range of dating issues and the trillions of dollars cost to continually update and adjust for scientific dating. January 1, 2000, was clearly not the only date issue to be resolved as time progressed.

The example shown in the box illustrates what might be done using only a single extra digit. For many date- and time-keeping purposes, it might be desirable to include not only century, year, month, and day information but also weeks, hours, minutes, and seconds. Thus if a date key is used to identify which format is being utilized, even the following 16-digit date format could be used if needed: x-yyyy-MM-ww-dd-hh-mm-ss.[29]

Possible Date Format Key Using One Additional Digit

Key	*Definition*
1	ISO date format with four digits for years (yyyy-mm-dd)
2	U.S. default date format with four digits for years (mm-dd-yyyy)
3	Normal European date format with two digits for years (dd-mm-yy)
4	Normal European date format with four digits for years (dd-mm-yyyy)
5	Normal U.S. date format with two digits for years (mm-dd-yy)
6	Julian date with two year digits (yy-ddd)
7	Julian date with four year digits (yyyy-ddd)
8	Astronomical time (dddddddd) starting from January 1, 4713 BC

As mentioned previously, public concerns and welfare pointed toward successfully supporting physical infrastructures as winter gripped the Northern Hemisphere. It was to everyone's benefit to have this global issue solved or at least patched to keep public utilities running to provide heat, water, lights, and power. January first in the northern regions is a vulnerable time for infrastructure failures, threatening lives on a massive scale.

Additionally, in the mid- to late 1990s, hackers were active primarily for financial gain, both personally and for state-sponsored political motives, with no evidence that rogue states were attempting to sabotage enemy nation-states involving Y2K remediation efforts. As a fortunate consequence, remedial work was not known to be sabotaged by political enemies.[30] The lack of state-sponsored internet attacks cannot be stressed too strongly; the same software issues today would surely have a completely different outcome with sophisticated sabotage interfering with remedial processes.

Nevertheless, the United States established the Center for the Year 2000 Strategic Abilities in Colorado Springs in early 1999, for global leaders to observe one another to "ensure no nation was launching a military offensive."[31] Just a single article in the thousands I have saved cautioned readers about information warfare, sabotage, or terrorism. Bruce Berkowitz in the *Wall Street Journal* referenced the outsourcing of critical Year 2000 compliance work to COBOL programmers overseas, since older computer languages were still widely used in developing nations and needed for many legacy computer systems. Any access to vulnerable mission or defense critical systems could be disastrous; "Criminals, terrorists and hostile governments will find opportunities in the [Y2K] confusion."[32]

Those heavily regulated banking institutions were considered particularly vulnerable to the safety and availability of cash to keep the country's economy going. The Federal Reserve was reported to distribute an unprecedented $50 billion additional cash to their member institutions, totaling $200 billion in circulation.[33] As an unintended consequence, return of that extraordinary excess cash to the Fed following January1, 2000 was not the best-kept secret in this nation. It was extraordinary that it went without incident, however.

And while we are talking about the Fed, former Federal Reserve Board Chairman Alan Greenspan noted in mid-1999 that he was "a bit of a computer nerd," designing a variety of software programs in the "early days of the computer revolution." Credited with the country's long economic expansion, Greenspan was testifying in support of the Y2K legislation limiting liability and damages to support economic stability. Mr. Greenspan said some of the programs he designed contain the Y2K bug, not intended "deliberately or maliciously," nor thought by programmers to still be in use in the late 1990s.[34] Again, this points to the generally held opinion that the "Y2K Bug" was an error of programming judgment, when in fact it was an economic imperative at the time.

Industry and manufacturing in the late twentieth-century global economy of supply chains and markets faced very real concerns for their partners' (also then called *counterparties*) successful Y2K compliance, giving rise to new contractual obligations and guarantees. Indeed, commerce struggled with both internal and external challenges. *Software Magazine*'s final Year 2000 issue in October 1998 offered an article maintaining that Y2K supply chain problems were of primary risk to business enterprises. The author, Matthew Schwartz, highlighted the Gartner Group's surveys citing South America, Asia, Africa, Eastern Europe, and the Middle

East as likely not being Y2K ready, as well as suppliers with fewer than 2,000 employees in countries with greater readiness expectations. Add government, infrastructure, and monetary institution compliance issues in those regions, and the risks become greater. Schwartz recommended that just-in-time manufacturers (which order parts and supplies only for immediate order fulfilment for the purpose of reducing inventory, management, and storage costs) consider stockpiling inventory in case of supply chain interruptions, even if costly.[35] None of my own research on Y2K supply chain issues has found any mention of ISO standards, which require identification of alternate suppliers for ISO certification.

In the same *Software Magazine* issue, William Ulrich, popularly known as "Dr. 2000," replied to Year 2000 questions from corporate management. Regarding responsibility for due diligence with mission-critical suppliers, Ulrich observed: "If business people handle these issues up front, you may determine a win-win situation. If lawyers are dealing with these problems at the back end, you can assume that there will be few winners."[36]

The psychological implications of the coming millennium change and how these affected both corporate and consumer responses to the challenges were being intensely felt.

Referencing Ramsay Raymond, who regarded us as a culture held hostage by fear of litigation, I agreed that corporate America was never more acutely aware of that now that it was facing Y2K challenges. Raymond's examination of the psychological pressures of the millennium change made it clear that the focus was moving from actual remediation efforts to corporate protection from liability and litigation. With litigation expenses from Year 2000 estimated to be a trillion to a trillion and a half dollars, anxiety over the cultural cost of this technological challenge seemed crippling.[37]

Along with commercial, industrial, and financial enterprises' focus on the coming millennium, religious institutions also got involved. The Unitarian Universalist denomination created the UUY2K Project with a website including related explanations, resources, sermons, a formidable bibliography, and "Psycho-social,

Spiritual & Ethical Considerations" (PSSE). "The way we think about ourselves and the progress we take for granted constitute a cultural religion. Y2K has the power to challenge some of our most basic assumptions." Noting that conservative religious interpretations of the "Wrath of God" may play a part in people's fears, the project was formed to explore the search for truth. PSSE included "The Y2K Connections Games," a card game of "What-if Scenarios" with the purpose of preparedness without panic.*

Capers Jones, technology professional, author, and Year 2000 expert, offered a very interesting analogy. Jones compared the divergent opinions of the magnitude of Y2K issues and failures to those leading up to the Civil War. The likelihood of secession and ultimately war following the election of Abraham Lincoln to the presidency met with widely conflicting opinions among congressional leaders in Washington. Many senators and congressmen thought a resulting war would be short-lived with a low rate of casualties, while others feared a more violent conflict. Of course, no one knew which view was accurate.[38]

An early 1999 feature article in *USA Today* examined growing Y2K jitters through a historical lens. "Virginia Tech professor Marshall Fishwick notes that when the year 1000 approached, people 'stood in the streets waiting for the end of the world.'" Fishwick concluded that current Y2K media hype would inevitably lead to similar reactions.[39]

Y2K consultants sprung up with greater frequency as the new millennium drew near. While panic might not be the operative term as midnight 1999 approached, certainly IT professionals were on call 24/7. Boards of directors became especially interested in successful remediation efforts, particularly in the more heavily regulated financial sectors. Insurers were doubly uneasy, both for their own legacy systems (often decades old), as well as for the invoking of potentially massive insurance proceeds for

---

* The UUY2K Project, headed by Reverend Dacia Reid, included a Y2K Connections card, website, telephone numbers, and email addresses.

clients' Y2K failures. *Mealey's Litigation Report* suggested that Y2K litigation would bear a $1 trillion price tag, much of it suggested to be covered by corporate insurance.[40] The U.S. insurance industry in the late 1990s had less than $400 billion in insurance reserves to pay claims, portentous of a cataclysmic collapse if litigation successfully relied on insurance proceeds. Law firms developed strategies while insurers formulated defenses for a risk neither contemplated nor premium rated.

JANUARY 1 AND BEYOND

January 1, 2000, arrived on a Saturday, with U.S. financial markets closed on Sunday. The following Monday, January 3, 2000, was also a holiday. Therefore, the world waited anxiously through the weekend and the holiday to see if finance and business would open as usual Tuesday morning, January 4, 2000. CNBC ran a full-page ad in the January 2, 2000, *Sunday New York Times* for "24 hours of market-to-market coverage" around the world.[41]

There is much to investigate and analyze within the major U.S. strategies, efforts, upsets, and conclusions of Y2K remediation efforts leading up to January 1, 2000, both temporary and permanent. While many Americans regarded Y2K as a "non-event," or termed by some "compatible self-interest," we will see throughout this book that Y2K was nevertheless the greatest cooperative effort ever undertaken by humanity.

Y2K AND ITS APPLICATIONS FOR GLOBAL CRISIS MANAGEMENT

Y2K offers us more than a fascinating historical account. Examining the business and governmental forces during the period preceding the new millennium offers insight to understand and plan for future global issues. Mapping the forces pre-millennium and factoring the current rise of rogue states can assist planners, whether for climate change, pandemic crises, supply chain issues, or huge technology steps such as the dynamics of artificial intelligence (AI), nanotechnology, and quantum computing.

Looking to future crises brings us inevitably to artificial intelligence, quantum computing, and other technological advances progressing faster than law, government, and the general population can grasp. As I write this book, AI in particular threatens the same dynamics of optimism and fear, hope and foreboding. Like Y2K, a limited number of people are responsible for AI's complex technology, yet their work will conceivably impact—and disrupt—every industry, as well as our daily lives. Our fears are justifiable; aided by the staggering power of quantum computing, no one working in or affected by AI knows the ultimate outcome.

Will a consistent roadmap of cooperative efforts from every sector successfully control the staggering power of AI? Could two and a half decades of technological advances in the dark side of cyberwarfare, greed, and control prevail? Will corporate and international leaders and public servants rein in and restrain AI's destructive capabilities?

Ultimately, those answers, dear readers, rest in our hands!

1

READY OR NOT

As the 1900s drew to a close, readiness issues for the new millennium grew. The opinions of a multitude of experts began to expand and vary as January 1, 2000, approached, fueling dramatic limits to the Y2K apocalypse spectrum. Survivalists were sure of Armageddon; ho-hummers were certain of a "non-event." Public and private sectors naturally pursued their own interests when staking a position on the matter. Corporate entities large and small worked diligently for date compliance while taking advantage of federal legislation for protection against litigation. Insurers, like all corporate neighbors, labored on internal readiness while pre-defending against coverage assertion by the tort bar. Tort attorneys, like all corporate neighbors, labored over internal readiness while maintaining a positive assertion of coverage for clients poised for litigation.

In November 1999, President Bill Clinton announced that the United States was "well on its way to being Y2K ready." According to the *New York Times*, Clinton predicted that the "Year 2000 computer problem would not disrupt the federal government or critical companies in the private sector, including those providing most of the nation's financial services, power, phones and transportation." The same article went on to quote experts cautioning against such optimism, including Clinton's Y2K Czar, John Koskinen, who remarked, "We are urging people not only to prepare, but to prepare early." A box insert accompanied the article, offering dismal assessments of readiness in areas such as 911 services, health care, education, and international preparedness.[1]

THE STATE OF COMPUTERS AND CODING AS THE NEW MILLENNIUM APPROACHED

The Y2K problem, when it was discovered at various times prior to and during the 1990s, was blamed on the shortsightedness of software programmers who used a two-digit date field. It is understandable that the cost of memory storage might be paramount, as all organizations, both public and private, faced the growth of their IT departments and the increasing costs of trained computer professionals. A quick look back at data entry costs over the last half of the twentieth century saw the lowest salaried women clerical forces being replaced by largely male-dominated, highly compensated computer programming professionals. I noted that fact decades ago, as well as suggested that for the first time in history corporate executives had neither the ability to step into any functions of their IT departments, nor knowledgably understand the costs.

IBM and the 80-column card were the standard for finance, banking, and insurance automation. Few, if any, authors of Y2K articles in the 1990s appeared to have been IT professionals (then known as information systems [IS] professionals), and unaware that most of the available 80 columns were heavily coded. In the latter half of the twentieth century, data storage cost considerations required just two or three digits to represent most complex and arcane financial data information.

In fact, dates were among the simplest of the coded fields. Two-digit years were not only enormously cost effective, but logical. This was no mistake or programmer oversight; this was pro forma for the digitized language of computing. Dates may easily be repeated thousands of times throughout a single program, representing thousands of dollars in memory expense in the middle of the last century. If there were any oversights to mention, cataloging where date fields appeared in software applications might be proposed. But that seems at best to be a stretch of early foresight.

In a well-researched feature article in the January 1999 issue of *Vanity Fair*, author Robert Sam Anson explained the scale of the problem: 1.2 trillion lines of questionable code globally, plus 30 billion microprocessors. He looked back to when the programming language, COBOL, was created by a team that included Grace Murray Hopper, an extraordinary and rare woman in the field of computers in mid-twentieth century. Known as "Amazing Grace," Hopper was the first female admiral in the U.S. Navy.*

COBOL, shorthand for **CO**mmon **B**usiness-**O**riented **L**anguage, was a versatile programming language suitable for a vast array of business applications for "the functioning of virtually every business computer." Those computers used the 80-column Hollerith card, compacting coded data, surprisingly, since 1890.[2] By the 1950s the 80-column card became more and more cumbersome, with many common programs needing a card deck to run. Magnetic tape commonly replaced many of the Hollerith-based programs, requiring the same banks of data-entry clerks to punch round instead of square holes in the tape. However, financial sector legacy systems often retained the 80-column card for

* I had the honor of meeting Grace Hopper when she met with our group of women systems designers in Boston in the 1980s. She gave each of us a short length of coated wire, representing the length of time sound travels in a nanosecond. I hope I still have it somewhere.

continued servicing of products no longer on the market but still with active accounts.

Anson went on to reference the background of the two-digit year field in critical terms. Among those Anson mentioned was Robert Bemer, inventor of the "escape" key, one of the creators of ASCII programming language, and a COBOL capability for either a two-digit or four-digit date field. Bemer's advocacy of a four-digit date year in 1960 was joined by forty-seven industry and government specialists seeking a universal four-digit date standard. By 1967, the National Bureau of Standards was directed by the White House to settle the matter, but ultimately the U.S. Department of Defense, "the biggest computer operator on earth," was unwavering in its two-digit date field position. Bemer was joined by ally Harry White, a DoD computer-code specialist, who had no more luck after transferring to the Standards Bureau. Pentagon pressure prevailed to continue the two-digit date field. Anson concluded with "Mindful of government contracts, big business went along."[3]

Anson compared mid-1960s data storage costs at $761 for the equivalent of a Danielle Steel novel, to "less than a thousandth of that" in January 1999, concluding "it was pound-foolish writ gigantic."[4] Those costs are not as easily quantifiable, which makes Anson's conclusion unpersuasive in this author's opinion.

Little noted in light of the tremendous attention to the Year 2000 date problem was an issue of 1999, where "99" was commonly used by mainframe operators as filler for an unknown or "other" data field, as well as an "end of file" designator. Thus, when "99" was entered into computers to represent an actual date field, system crashes were commonly reported to have occurred. Clearly this was not as large an issue as the "00" date, but I am aware that "99" is still used in entries for required fields when the data is unknown to accommodate moving forward.

January 1999 news reports indicated some random "99" issues were easily repaired, many reportedly with the Year 2000 programming fixes. Issues with September 9, 1999 designated as

"9999" or "9/9/99" had also been noted as causing the same issue with the added consequence of the deletion of huge quantities of data. Complicating both the 1999 and Y2K remediation issues were the enormous technology resources committed to the European euro currency, which began its launch on January 1, 1999.

HOW TECHNOLOGY DEPARTMENTS AND SOFTWARE CODERS RESPONDED

It is clear that much of the popular press was cavalier in their opinion of two-digit versus four-digit year computer usage. Over the years I have opined that technology departments were the first corporate divisions that the chief executive could not operate. This shift was particularly palpable as the new millennium approached, since the complexities of IT were commonly beyond the capabilities of the founder, the product entrepreneur, even as the costs of IT became a greater and greater budget segment. Highly paid IT professionals replaced rows of low-wage data-entry clerks. Systems specialists handled more and more complex data sets and processing.

I knew from my experience working during the 1970s and 1980s systems era that programmers somewhat routinely wrote code that was unnecessarily complex or arcane. It was even common to create modules with program erase instructions after a specific time failing the programmer's login. Both were intended for either job protection or firing retribution, and were not the best-kept secrets to management. Anson's article referenced the same "COBOL Cowboys" in the 1990s coding hidden dates or using arcane terms for date sequences.[5] We can conclude that the task of reviewing such code by anyone but the original programmer was Sisyphean.

Perhaps singularly unique during this technological event, the downsizing of older, experienced COBOL programmers suddenly emerged as shortsighted as it became obvious that Y2K compliance issues primarily affected legacy systems programmed with COBOL. IT units were strained by review of every internal and external computerized system. Long-retired COBOL programmers were lured out of retirement to assist with legions of legacy systems while veteran programmers still employed were tempted by lucrative fees from former employers desperate for experienced talent. December 31, 1999, was most likely the first immovable systems deadline any IT department ever faced.

Nevertheless, even with experienced COBOL programmers working, systems at that time were a series of patches and updates, rarely carefully documented. Dr. Leon A. Kappelman, cochair of the Society for Information Management Year 2000 Working Group, said, "It's as if you were building a bridge and let every riveter pick their own kind of rivet and drill different holes and use different rivet guns."[6] The result was "Spaghetti Code," a term widely used for at least three decades of the twentieth century, describing a warren of patched code to preserve legacy systems until the useful life of their product ended and computer support was no longer needed.

Date fields of MMYY or YYMM for the Year 2000 and any year YY≤12 could be misread.

Pundits were counseling IS professionals that Y2K problems were catastrophic, and to consider retirement quickly. Michael Cohn, an IS professional in Atlanta, satirically suggested IS professionals' retirement, resignation, re-engineering (replacing all existing IS staff with new), repairing, recycling time (start the calendar at Bill Gates birthday, dubbing it G.A.G. with cautions about resultant Windows problems), relocating (moving your data center to Honolulu to buy a couple extra hours), and finally, relaxing—we're doomed anyway! Cohn's satirical observation had a semblance of genuine fears as the magnitude of the challenges became clear.

HOW THE U.S. GOVERNMENT RESPONDED

Those in and around technology weren't alone in their concerns. As the millennium approached, *Government Executive* reported in mid-1996:

> IBM has announced that its S/370 line of mainframes cannot be modified for the Year 2000. And some Unisys machines can't either. In addition, the basic input-output systems (BIOS) on older PCs will not be able to accommodate the century change. As a result, some machines will require a date-command execution every time they are booted up. Recent tests resulted in a 97 percent failure.[7]

Y2K anxiety began to increase!

Leading into the new millennium, the United States had striking vulnerabilities. The Department of Defense then and now gathers more data using the greatest amount of computer power on the face of the earth. Steven Levy, in an article for *Newsweek* in mid-1996, quoted Bob Molter, the DoD's Year 2000 official who commented that he did not think nukes would be launched, but said "without fixes, our sensors might regard, say, troop movement in Chechnya as occurring in the McKinley administration." Levy went on to compare the Millennium Bug to the technology equivalent of the savings and loan crash: "Like the latter, the 2000 problem came seemingly out of nowhere, a huge bill slapped on our doorstep. And like the S&L fiasco, the discovery has occurred long after the perps have flown the coop." Quoting Peter de Jager, Levy noted that "It was an error of procrastination, a conspiracy of compromise."

Levy continued:

> The bill was going to come due, but it would be someone else's bill; the guilty parties figured they'd be somewhere else when the bits hit the fan. And they were right. When the hideous bill for this fix arrives, computer haters will have their field day. But it's a bad rap. The real

> culprits have themselves been around for millennia: the all too human foibles of denial, short sightedness and greed.[8]

"This is a culture held hostage by fear of litigation," said Ramsay Raymond sagely in July 1999. Raymond, a psychologist and owner of The Dreamwheel, a Concord, Massachusetts, advisory firm, concentrated on the psychological pressures of Year 2000. Never more acutely was corporate America aware of this fear than when facing Y2K challenges. More and more, focus was shifting from actual remediation efforts to corporate protection from liability and litigation.[9]

At the time, I offered my own support to Raymond:

> With the American bar estimating litigation expenses from Year 2000 to be a trillion to a trillion and a half dollars, anxiety over the cultural cost of this technological challenge seemed to appear crippling. Y2K litigators were expecting insurance proceeds from the casualty/property sector to offer significant funding of those expenses and fees with extraordinarily imaginative arguments for coverage. Insurance executives fiercely defended against proceeds for Y2K software adjustments—with extraordinarily imaginative arguments defending against coverage. The anticipated litigation fees were almost two and a half times the existing U.S. insurance reserves, signaling the possibility of a collapse of the entire U.S. property and liability insurance sector.

Mid-1999 was a time of increasing uncertainty and anxiety as the costs of systems analysis, review, and remediation and the anticipated costs of litigation and defense were steadily increasing.

Raymond expressed an unusual sociological view, unhampered by self-interest or fears of those whose Y2K ponies were in the race. Her words are important to include here in order to present the perspective of those working within the human side of this and every such global crisis issue:

> It is my position that unchecked litigation with financial awards that far exceed just recompense is actually damaging to the social contract.

> First, it breaks the fabric of trust that is essential to the common weal: Secondly, it allows greed to motivate litigation, which in such an instance as Y2K could cause a veritable feeding frenzy; and Thirdly, it passes on huge, largely invisible costs to the average citizen which drive up the cost of living and redirect monies towards ends that have not been consciously chosen by those spending it. And Fourthly, the emphasis on the part rather than the whole may be our undoing—e.g., the emphasis on individual rights and corporate rights while noble in itself, when out of balance may obscure the greater needs of society as a whole. The thinking remains localized. If anything, Y2K is teaching us that things are not just local and never have been, though the effects are local, it is huge systems, like the ecosystem that govern what happens. We must learn to think within the context of the whole and of future generations if we can be considered a truly sophisticated society, The native peoples of this land have a tradition, that all decisions taken by the adult community must be made in light of its effect on those living seven generations in the future. This is wisdom indeed. It is our failure to do so that has created the technological problem (the Year 2000 will never get here, so thinking was in the last half of the century) and if we fail to do so in how we handle the problem we may become deservedly stuck in the mud of our origins.[10]

Raymond prepared testimony to the U.S. Senate for their "Senatorial Inquiry into Neighborhood Preparedness for Y2K Disruptions." She offered a societal perspective, which gave me pause to reflect on the fundamental motivations of the primary actors on the Year 2000 stage. Government bodies seemed almost entirely litigation-focused as they enacted both self and business protection legislation; plaintiffs' attorneys were completely focused on litigation opportunity with insurance as the funding mechanism; the insurance sector concentrated on defenses against invoking coverage for Y2K failures. The technology sector, charged with finding and fixing noncompliant code, was focused on the task, while corporate counsel drafted contractual language to avoid litigation.

It is worth noting some of Raymond's comments here for the purpose of understanding that our government officials in Washington heard her important words in mid-1999.

> I will limit my comments to two areas of concern.
>
> I. Paralysis of Public Leadership Because of Pervasive Fear of Litigation
> The computer-generated problem clearly belongs to a category of social disruption that is unprecedented in nature and scope, and unfamiliar to all. That being so, emergency plans might be instituted based on what actions—or non-actions—will be for the highest good of all.
>
> One proposal would be that the senate suspends citizen's rights to sue anyone for problems stemming from Y2K computer error, that Congress pass temporary laws modeled on the ones recently instated in Britain and France which have rendered it impossible for individuals or corporations to sue for Y2K instigated damages. (Nancy James, an insurance broker active with the Chamber of Commerce on Y2K issues in Concord, is knowledgeable on the subject, has written articles, and would be a useful resource on the subject: E-mail: npjames@compuserve.com.)
>
> II. Mental Preparedness. The Need for National Leadership
> The lack of highly visible leadership on the town and city and state and federal levels I believe are hampering the ability of citizen groups such as ours to invite others to take the problem seriously—even though the Red Cross and other federal agencies have gone public on the matter.
>
> Quality leadership at the national level can make a very real difference in mobilizing national attention, ingenuity, and resourcefulness. The decision to avoid panicking the public by avoiding the issue in public postpones the inevitable and seriously aggravates it to have people discover the problem when it happens is a clear set up for social mayhem and psychological fragmentation.[11]

In response to such pleas, while existing bars to litigation brought against federal, state and local government bodies offered protection, some states enacted specific Year 2000 immunity (see chapter 6). The possible catastrophic consequences of infrastructure failures mid-winter could not easily be held against those responsible federal, state or local municipalities. Oh, the irony.

At this juncture in Year 2000 understanding, compliance analysis, planning, and testing, anxiety and uncertainty were high, very high as competing forces planned both defense and offense. While it inevitably might have been difficult to perceive, let alone believe in cooperative effort as a unifying principle, it is evident from my research that the greater good was the fundamental objective.

2

TECHNICAL SOFTWARE SOLUTIONS TO Y2K

The year 1999 was filled with news reporting on Year 2000 compliance concerns in every sector. Corporate America was anticipating issues with the beginning of fiscal year 1999 on customary April 1, July 1, and September 1 dates.[1] Estimates at the time were that one-tenth of 1 percent (.001) of embedded chips would fail. That is 25 million, minimum, of the 25–50 billion estimated embedded chips.

Statistics were published widely regarding anticipated failure rates:

200,000 mainframe computers at 50% failure rate = 100,000 failures
400,000,000 PCs at 5% failure rate = 20,000,000 failures
25 billion embedded processors at .1% failure rate = 25,000,000 failures[2]

Nationally recognized Year 2000 expert, consultant, and author Capers Jones, in his March 9, 1998, "Abstract: Dangerous Dates for Software Applications," concisely places the above statistics in context:

> Incidentally, in the entire 50 years of the software industry there has almost never been a major software application released to users where 100% of the latent errors were found prior to deployment. The current U.S. average overall is about 85% of defects are removed and 15% get deployed. There are roughly 36,000,000 applications running in the world which have Year 2000 date problems in them. It is very naïve to think that 100% of these will be repaired in time. It is also naïve to think that for any specific application that 100% of the Year 2000 date references will be found and repaired.[3]

Jones references routine systems applications functions: program, test, debug, deploy. Repeat. As Jones points out, it is very rare for an automated application to be bug-free when implemented. These staggering estimates sum up the magnitude of Y2K compliance remediation work necessary, which helps to explain the intensity and urgency as 2000 approached.

Canadian computer consultant Peter de Jager was widely credited with bringing attention to Year 2000 with his essay, "Doomsday 2000," published in *ComputerWorld* in 1993. *Scientific American* published a de Jager article in January 1999, describing details of various Y2K software solutions. De Jager outlined problems with incompatible versions of upgraded compilers, lost software source code, missing documentation, and various date configuration schemes, alerting readers to possible deadly consequences of date confusion errors. Parenthetically, de Jager likened recreating such lost code as "a heinous process that has been compared to retrieving a pig from sausage."[4]

Software management expert Professor Leon A. Kappelman of the University of North Texas offered insightful comments: "It's typical information technology," and "You don't get any recognition until the last 30 days that your project is going to be late."[5] This suggests that successful Y2K compliance by 01/01/2000 would not be known until November 30, 1999. A rather chilling but realistic observation.

But talented technology professionals were designing innovative solutions to the specific problem of January 1, 2000—which was not applicable for a day-late, module-short negotiation, as was (and still is) typical for most system applications deployments.

INNOVATIVE SOLUTIONS TO Y2K DATE COMPLIANCE FOR 01/01/2000

A number of possible solutions were posed to the Year 2000 date change issue. Volumes of technical advice and summaries have been written on the many software theories and solutions for Year 2000 remediation. That was not the work with which I dealt; I have not worked in systems since 1982. However, I will offer an overview of those interesting and sometimes arcane compliance solutions detailed in the thousands of documents I have accumulated during that time period.

Interestingly, one of the best summaries of various solutions was carried in the Summer 1998 issue of an insurance periodical, *CPCU Journal*.[6] Many technical treatises have been written on solutions to the Y2K date problem; "field expansion" (expanding the existing two-digit year date to a four-digit date) being identified initially as the most common. The issues and pitfalls identified in field expansion presented a host of complexities due to issues of differences of date notations within the same software program, later versions of partial program and module updates, and various application integrations, both known and unknown. In my introduction I listed various date configurations as published by Capers Jones. All to be discovered and analyzed, fixed and tested.

CPCU Journal also explained the "encapsulation" and "bridges" methods. Regarding encapsulation: "This method involves subtracting a constant from the two-digit year for processing purposes and adding the constant back after the processing is complete. Typically, 28 (or a multiple of 28) is chosen as the constant because the date-to-day relationship in the calendar repeats every

28 years." It was noted in the article that the insurance industry relies heavily on historic data, making encapsulation not reasonably feasible. Bridges involved creating duplicate file modules, expanding the date fields, and bridging the module back into the program for processing. This method was considered less disruptive to non-date-sensitive programs, while being inadvisable for programs needing real-time access.[7]

Among two of the more imaginative methods cited in many publications, which both appear to be triage models, were the "sunset" method and the "windowing" or "sliding window method." The sliding window method set software instructions only for certain dates below a certain threshold for twenty-first-century dates. A common window was 10, giving software programmers an extra ten years to find, replace, or reprogram non-Y2K-compliant code. Patch code for two-digit dates 01-10 would be interpreted as 2001 to 2010, with 11 up as 1911 onward. Fidelity was likely using 10 as a window, but even with ten years for ongoing Year 2000 issue identification for remediation, one very large financial module had been neglected and crashed.

Capers Jones's "Abstract: Dangerous Dates for Software Applications" also mentioned windowing. As he explained,

> The windowing method establishes a fixed interval time period, such as 1915 to 2014, and uses external program logic to deal with all dates within that period. Assume that your window runs from 1915 to 2014. Dates below the midpoint of the window such as 03 or 10 are assigned to the 21st century as 2003 or 2010, while dates above the midpoint such as 97 or 98 are assigned to the 20th century as 1997 and 1998. Windowing for a portfolio can be finished in roughly 18 calendar months, so this has become one of the popular methods with late starters. However, windowing exacts a performance penalty and assumes that everyone using the data or the application knows about the existence of the windowing routines.[8]

The other triage model, sunset, recommended using an application until it became obsolete, then replacing it with Y2K-compliant software. Detailed accounts of sunset usage are rare, since it does not in any way address the failure rate for noncompliant systems, and even then seemed to be the antithesis of needed remediation efforts. Sliding window method may have been a most common backroom strategy for many enterprises needing additional time for review and updating of multiple systems and software forms. The danger in this piecemeal approach lies in continuous dashes to the finish line as systems, particularly in the financial sector, are scheduled to be run after 01/01/2000. In addition, both sunset and sliding window run the risks of interconnectivity, or linking, with other systems, which may have been missed during triage. In fact, those desperation choices were associated with the seven stages of grief: shock, denial, anger, bargaining, depression, testing, and acceptance.

Both scanning browsers and clock simulators were used to test for Year 2000 vulnerabilities. A third approach, compression technology, was used to impose four-digit dates into two-digit date fields. It was not only complicated but required conformity across programs, an improbable condition.

It is evident from the brief descriptions of a few of Y2K remediation choices that even opting for a method was complex, irrespective of the complexities of executing the actual process successfully. How companies determined which method to commit to, along with any bumps along the way, were understandably not publicized.

Among the solutions ultimately found useful, there were some complex, arcane, and often flawed Y2K repair mechanisms in competition. For example, one very intriguing solution reported by the *New York Times* in mid-November 1999 involved a fix developed by Willard H. Wattenburg called the A0 system. Wattenburg, who founded Y2K-OK, LLC, developed a hexadecimal date configuration in conjunction with IBM and Allstate. The application uses base 16 numbering system, which uses alpha characters A to F for numbers 10 to 16. This A0 hexadecimal system was purported to delay Year 2000 turnover to "00" for fifty-nine years. Not only did the Wattenburg A0 system have its own internal problems, but issues in communicating with non-hex systems prevented its adoption.[9]

Software Magazine takes credit for the first (and they said, the best) creation of a timeline managing Year 2000 projects, "Year 2000 Survival Guide," published regularly from mid-1996. In the last of the Survival Guide issues, October 1998, editor John Kerr noted in his letter to readers that buried in a Gartner Group's summer 1998 report was that 50 percent of all companies were *not* planning to test their Year 2000 compliance code. Kerr added a quote from legal expert Steven Hock: "If there's one thing juries understand, it's testing—including testing of all the products you buy."[10]

Credible accounts such as the above had to leave readers bewildered, even troubled. Fifty percent of companies were *not* planning on testing? Hock's conclusion about juries was at the core of wondering how such decisions to not test code could have been made. Was it a misprint? Strange or incongruous corporate posturing? What we did not know then was how reliable the untested code was as the new millennium arrived. What we did know is that few reports of massive Y2K failures were reported, either from corporate officials, regulators, or angry consumers.

As noted in my introduction, many hoped for some "silver bullet" solution, a system-wide debugger, which would eliminate the need for review and recoding of every line of software code. My cartoon of a silver bullet expressed my opinion, and most if not all experts' opinions, of some magical Hail Mary pass solutions to Y2K.

THE YEAR2000 POLICE BOTHER YOU ON THE GOLF COURSE

Courtesy of the author.

Solutions were anything but! It was the hours, the hard work, and the dedication of computer and technology professionals that accounted for success.

While the ten-year delaying tactic (windowing: offloading 1900–1910 documents onto a separate server and rolling back all dates ten years, offering ten years more for remediation or replacement) was useful, it was used only as a fallback for most remediators for lower sector priorities when time was running out.

While there were volumes written at the time on solutions for 01012000 software date compliance, these brief outlines of the more common and interesting methods serve only as a quick refresher. The complexities inherent in each of these solutions belong within the highly trained professional software technology sector. My review provides merely a snapshot of what those IT professionals dealt with.

ADVISORY AND MANAGEMENT CONSULTING SOLUTIONS

Most, if not all, technology consulting firms necessarily were involved in Y2K software advisory and programming services. Some provided on-site management, project, and personnel, while others offered toolkits and manuals for compliance projects and benchmarking. There were numerous competitors in the Year 2000 blueprinting and strategic planning sector, offering a vast and varied range of services. In addition, outlined briefly below, were all-inclusive guides, certification, and evaluation sources—an overwhelming breadth of ancillary if not distracting options.

Keane, Inc., confidently offered its Y2K remediation services: "As one of the largest software services firms in North America, Keane had the financial strength and resources to help clients minimize risk and successfully achieve Year 2000 compliance" and "proven approach with full project accountability." Keane offered a three-phased approach: enterprise planning, strategy development and confirmation, and implementation.[11] The global technology giant SAP predictably also offered Year 2000 compliance services, but with considerably fewer assurances of risk mitigation.[12]

Interestingly, one of the most closely watched early Y2K litigation cases was against Keane, brought by Pineville Community Hospital Association in Pineville, Kentucky. In 1998, Keane acquired Source Data Systems, which had a mid-1995 contract with Pineville for equipment and software. The dispute involved Pineville's Year 2000 compliance costs. (This case is covered in detail in chapter 5.)

Among the thousands of documents saved to the run-up and following the millennium change was a letter and "RoadMap 2000" brochure from PKF Technologies accountants sent to me. The letter, interestingly, was not dated, nor did it carry any location address, while including references to predictions to a 70 percent chance of economic downturn, $600 billion estimated remediation costs, and $1 trillion litigation expenses. The brochure mentioned "software," but pertaining only to a "Y2K Forms Toolkit" offering a management guide to administering Y2K compliance, implementation, legal, insurance, accounting, financial, and documentation projects.[13] I'm not sure how I might have been mailed such an offer, but I regard it as a good, if puzzling, example of the marketing opportunities the new millennium posed to management.

Digital Equipment Corporation of Maynard, Massachusetts, (often referred to as DEC) published a document folder, "Year-2000 Solutions Digital Equipment Corp." on December 10, 1996 with detailed explanations of the Y2K software issues, solutions, Digital's products, layered software products, and specifics of Digital's product compliance. Offering customers step-by-step compliance planning advice, the document included company personnel names and telephone numbers. Interestingly, in its introductory pages DEC offered real-world reasons why Y2K remediation was slow to be addressed, including "Annual short-term business objectives ignore long-term issues such as Year 2000 compliance" and "People resources required to develop and implement Year 2000 plans are already committed to other programs and projects." Promising Year 2000 compliance by the end of 1997 for operating system software and layered products, assistance services to DEC clients was offered for complete client solutions. A resource website listing surprisingly included IBM, one of DEC's fiercest competitors. Impressive as the portfolio was, the firm's legal department most likely insisted on the following caveat at the conclusion of the work: "Digital believes the information in this publication is accurate as of its publication date: such information is subject to change without notice. Digital is not responsible for any inadvertent errors."[14]

Well, we have to ask, who *is* responsible?

London-based law firm Withers offered a wide range of Year 2000 advisory services including identifying liabilities and legal risks, compliance program review, supplier negotiations, insurance review, IT employee recruitment, and response strategy services. Early legal conferences including many U.S.-based multinationals as well as U.S. and EU media press coverage articles and commentary were listed on the firm's portfolio credentials.[15]

Additionally, in those documents I saved during the late 1990s is a directory of "Year 2000 solution providers" prepared and published by the thirty-five-year-old prestigious Information Technology Association of America (ITAA). In 1996, ITAA boasting 9,000 members, offered an extensive 11-part questionnaire taking four to six weeks to complete. The association also offered an "ITAA*2000 Certification Program" with a catalogue of forty-eight Y2K service providers in its fourth edition (Spring 1997) and eleven more in its May 1997 addendum.[16]

Other opportunities for lucrative solutions came from organizations offering certification programs. The Software Productivity Consortium undertook a three-week evaluation for certification, with those successfully certified being listed on ITAA's webpage.[17]

Critics suggested the questionnaire and evaluation time might better be spent on remediation efforts.

While that may be interesting to some researchers, I do not offer specifics here, but think it important to note those opportunities that Year 2000 compliance demands offered to technology-based enterprises.

Part of the difficulty in faithfully reporting IT efforts in the leadup to the new millennium is that organizations across industries were understandably motivated to keep the details of their Y2K remediation labors private and away from regulatory and public scrutiny. What we do know is that those efforts resulted in an almost seamless transition to the twenty-first century with no apparent peril to human safety or security. Much of the nation and the world breathed a sigh of relief, went back to their lives, declaring Y2K a "non-event." After all, I suppose that's what success should be called.

3

CORPORATE PREPAREDNESS, BANKS, AND INVESTING

A Brief Look at How Corporate America Prepared

Much has been published about corporate America's efforts to make their computer systems Year 2000 compliant, at least from what corporations were willing to make public. Some of those strategies are interesting to recount to emphasize the magnitude of these projects. Corporations who hoped to achieve Y2K compliance had to make significant investments of time, money, and people. It was important not just for corporations to prepare as 2000 approached, but also to have people and strategies ready and on call for any unforeseen issues that emerged in the first few days after the clock struck midnight.

An interesting tension was present, in that everyone was talking about Y2K, and simultaneously few corporations wanted to disclose the details of their own progress, as complete transparency could potentially have a negative impact on public perception and the flow of business.

Insurance being the focus of my professional career, many of my records from that time are from finance and insurance journals expressing foreboding on two fronts: corporate internal Y2K remediation as well as insurance coverage, claims defense, and possible judgment costs.

I will first note *Newsweek*'s Steven Levy's commentary in June 1996: "If you've ever felt that computers were a scourge upon the earth, just wait three and a half years or so." Levy continues: "Think of the Millennium Bug as the high-tech equivalent of the savings and loan crisis. Like the latter, the 2000 problem came seemingly out of nowhere, a huge bill slapped on our doorstep. And like the S&L fiasco, the discovery has occurred long after the perps have flown the coop." Levy then quoted Peter de Jager, a Millennium Bug guru: "Management said 'it's years away, we'll fix it later.' It is an error in procrastination, a conspiracy of compromise."[1]

Corporate attitudes toward the issue were distinctly not uniform, however. The London-based *Economist* quoted David Starr, chief information officer of *Reader's Digest*, in late October 1997 as saying that Y2K was "the biggest fraud perpetrated by consultants on the business community since re-engineering," noting that his remediation costs were less than 5 percent of his IT budget. The article colorfully concluded with "many will be disappointed to reflect that the final years of the 20th century were spent paying for the sins of computing's callow youth."[2]

In these two perspectives, Levy characterizes Y2K as equivalent to the "S&L fiasco," while Starr describes it as the "biggest fraud perpetrated by (Y2K) consultants." Both appear to be unhappily surprised and irritated by the Y2K date issue, but both are wrong in suggesting that it was either a deliberate deception (Levy) or a consultant's deceit (Starr). Certainly, consultants had a windfall in Y2K remediation fees, but remember, they were almost as much in the dark as their clients in terms of what the outcome would be. They, too, had their own houses to remediate.

On the investment brokerage side, Y2K tests, sponsored by the Securities Industry Association, were held in mid-1998 to assist anyone trading stocks, options, or corporate and municipal bonds. Supervised by Coopers & Lybrand, the tests dealt with a small set of specific transactions, although even a single trade could involve forty to fifty steps. More complex issues of managing dividends, margin trading, and client account management testing were planned for the spring of 1999.

An interesting issue of *National Underwriter* in January 1999 outlined the "war rooms" insurers had dedicated for Year 2000 remediation and monitoring right through New Year's Eve, 12/31/1999, and the weekend. A photograph of Prudential's war room accompanied an account of the company's readiness planning. Travelers Insurance Company's Y2K facility in Hartford, Connecticut, was described as more of a "bat cave." Travelers, as with all insurers and national financial institutions, identified "two types of Y2K failures— site-specific failures and business-function failures," which of course included all location facilities in the country, as well as operational business continuity.[3]

It was rumored in Boston that Fidelity had reserved *all* of Boston's downtown hotel rooms on New Year's Eve 1999 for their IT employees. However, a *Boston Sunday Globe* front-page newspaper report for January 31, 1999, included BankBoston as planning to have staff on site through New Year's Eve, with both Fleet Financial Group and State Street Corp. banks reserving blocks of Boston hotel rooms. The same article published charts by industry category, and listed both Fidelity and Fleet Financial as being 100 percent compliant as of that date.[4] The Massachusetts Port Authority reported that 350 personnel would be working through the new year with Boston hotel rooms reserved.[5]

San Francisco–based McKesson HBOC, Inc., the nation's largest drug distributor at the end of the twentieth century, had a command post near Sacramento, California, according to a mid-1999 *Boston Sunday Globe* front-page article on the preparedness of America's healthcare services. Distribution of medications would be monitored constantly during the date change, with quarterly follow-up tracking of drug supplies.[6]

Acquired by Compaq in June 1998, Digital Equipment Corporation (DEC) had been the developer of OpenVMS, a widely used operating system supporting numerous industries from the stock exchange and financial sector to healthcare and manufacturing. Known briefly as VMS, it was critical to application users that the operating system be Y2K compliant. DEC's Nashua, New Hampshire, facility had a rapid response team on site for the entire weekend running through January 1 to troubleshoot any Y2K software issues. It was reported that the team experienced a very quiet weekend without issue.[7]

Various references to Y2K compliance estimates and cost comparisons were repeated during the run-up to the new millennium. MITRE Corp. estimated conversions for its weapons systems could be eight times more expensive than Y2K projects at civilian agencies due to requirements for new microchips. The Department of Defense's Y2K remediation estimates in mid-1996 exceeded its entire information technology budget. The Federal Reserve prudently began replacing legacy systems in the very early 1990s as their corporate solution in lieu of reviewing and patching existing systems for Year 2000.

Flashing back: Thomas Lamont of J. P. Morgan, addressing reporters on October 24, 1929, noted, "There seems to be some distress selling on the stock exchange." That date is known as Black Thursday; Lamont's remark was a vast understatement, comparable in 1996 to remarking on possible cataclysmic consequences of Y2K.[8] It was recommended in the mid-1990s to prioritize systems vital and those not so vital for Year 2000 remediation before January 1, 2000.

A brief January 1998 article in the *Boston Globe* referenced a Y2K software issue that shut down computers of First Call Corp., a Wall Street earnings estimates service provider and a unit of Toronto-based Thomson Corp. The shutdown prevented investors, bankers, and analysts from receiving earning information, statistical ratios, and analyst recommendations for January 1, 2000, and after. Credit Suisse First Boston declared the shutdown a "wake up call."[9]

Klear Solutions software reseller CEO Anne Brennan was quoted in *Women's Business* in mid-1999 as saying, "You'd have to be the village idiot not to make money this year." She predicted that based on the considerable spike in pre–Year 2000 compliance conversions, post–January 1, 2000, software sales would continue to be robust. Like so many in the Y2K remediation sector, her biggest challenge was finding qualified staff to fill customer demand and hiring and training experienced people from related business sectors. Brennan noted that manufacturing and small heavy industrials had narrow margins that did not offer the financial capacity for Year 2000 systems replacement, so she was expanding to IS (Information Systems) consulting to supplement new software sales. Nevertheless, Brennan predicted a healthy group of businesses waiting through 2000 to convert from legacy systems to new software.[10]

Yet, in fact, following January 1, 2000, computer and software sales lagged. In my opinion, this can be attributed to the residual problems with Year 2000 replacement hardware and software, as both had many problems requiring fixes.

RESPONSE OF BANKS: CASH AND Y2K ETHICAL ISSUES FACING BANKS

Bank consultant Alex Sheshunoff called the Year 2000 problem "probably the most thankless tedious task in the history of computers," and the Gartner Group "the biggest single information project the world has faced."[11] I addressed the American Bar Association in 1997 regarding the exposures, risks, and liability assumed by banks during Y2K. Banking in the United States has the most complex levels of constituents: federal and state regulators, boards of directors, depositors, borrowers, and investors. The primary liability is attributed to the directors and officers, where the literal buck stops for any financial irregularities. Year 2000 computer compliance presented an enormous challenge, and affected every American's financial stability.

Year 2000 compliance in the banking sector was considered to be at the highest readiness level due to recognition that our national monetary system had to continue operating in an uninterrupted manner through the date change. Nevertheless, in late 1998, Massachusetts state banking regulators proposed legislation declaring banks that failed mission-critical tests could be publicly declared "unsafe and unsound." The Massachusetts Bankers Association and other trade groups supported the regulation, though they doubted it would need to be used.[12]

As reported by the *Boston Globe* in early 1999, the availability of cash on January 1, 2000, was worrying depositors. Public confidence in banks was troubling local community bankers, their regulators, executives, and trade groups. *Globe* journalist Lynnley Browning noted that "only 10 percent of the $4.5 trillion in U.S. bank accounts is available in cash at any given moment." Banks carrying extra cash as the new millennium approached would consequently lose lending revenue, as the costs to insure the additional cash created an additional burden. Concerns of power outages affecting bank vault security necessitated careful monitoring of customers' accounts as well as requirements by regulators for paper backups approaching January 1, 2000.[13]

Browning interviewed Cathy Minehan, president and CEO of Federal Reserve Bank of Boston, and reported that the Fed would have "an extra $50 billion in reserves to bolster the $200 [billion] or so used daily in the United States" to accommodate the anticipated demand for more cash. Banks estimated that between 10 and 30 percent more cash would be needed, with the interesting consequence of banks having to sell assets to provide the cash; the significance of that would be shrinking assets and possible loss of consumer confidence.[14]

Bank issues were identified in another early 1999 article on the proposed Year 2000 liability limit legislation introduced in Congress. Claims against bank trust managers for lowered portfolio value and costly trade interruptions, as well as vendor and affiliated bank compliance issues due to Y2K problems, all were alleged to be possible. The list seemed limitless and provided more proof of why I picked banks for my early writing and speaking on Y2K exposure and risk.

However, consumer confidence should have been bolstered by the same report stating that 97 percent of the nation's more than 11,000 FDIC insured banks reported having spent $80 billion on Y2K remediation. In addition, FDIC member banks had long insured individual deposits to $100,000.[15]

The Fed's Minehan was also interviewed in an extensive article in the *Women's Business*, August 1999 issue, stating, "There is no need to panic." Minehan chaired the Federal Reserve System's Financial Services Policy Committee and cochaired the Federal Reserve System's Century Date Change Council, noting that the Federal Reserve transfers about $2 trillion daily to banks and to the U.S. Treasury. She stated that the risk of depositors making large cash withdrawals in advance of the millennium change was much greater than the risk of Y2K computer issues. Interestingly, specifics of the $50 billion additional reserves were not mentioned in this article.[16]

The FDIC published an undated consumer brochure briefly outlining Year 2000 issues and steps that member banks and financial institutions were taking to protect depositors' and borrowers' accounts. An important part of the brochure was a list of the four federal agencies supervising banks and savings associations, including postal addresses, telephone numbers, and web addresses: Federal Deposit Insurance Corporation, Office of the Controller of the Currency, Office of Thrift Supervision, and Board of Governors of the Federal Reserve System.[17] The Federal Financial Institutions Examination Council (FFIEC), which oversees reporting, standards, and examination of numerous federal banking agencies, had issued specific guidelines in December 1997. The FFIEC outlined the penalties for institutions not taking the necessary steps to be Year 2000 compliant.[18] This is only a snapshot of banking regulators' actions leading up to January 1, 2000.

Citibank enclosed a flyer with their statements to customers with an advisory to "Keep statements and records of transactions from the end of the year through the first couple of months of the Year 2000." This included a cautionary note with an assurance of confidence in Citibank's Year 2000 readiness: "We have not identified any major Y2K issues that may affect your Citibank account."[19]

Having saved that notice and highlighted the quote above, I will take note here of Citibank's openness to expressing what was obviously the situation of most, if not all, finance-based enterprises. Twenty-plus years later it still impresses me.

Much concern, caution, and advice offered to and about banks was greatly heating up during 1999. Ty Sagalow, chief underwriting officer at National Union Fire Insurance Company (under Chubb) warned of a range of issues banks faced, including federal securities laws, common law negligence and fraud, as well as fidelity and contract law. He suggested that criminal and administrative proceedings, as well as government investigations, were possible, in addition to issues of reputational damage and media and customer relations. In an article via NewsEdge, Sagalow identified three insurance underwriter responses to Y2K: 1) policy cancellation, nonrenewal, or the addition of Y2K exclusionary endorsements; 2) insurers remain silent on policy language regarding Year 2000, to later deny coverage for claims; and 3) underwrite and price the exposure for each client. Concerning directors and officers (D&O) policies, Sagalow concluded that, even with no Y2K exclusionary policy language, "D&O coverage probably won't be adequate for a Y2K."[20]

It was considered prudent for banks to have an unusually large amount of cash available for customers immediately leading into the new millennium. The public was warned that if credit- and debit-card machines failed due to Y2K glitches, cash would be the only alternative. The Associated Press reported on January 4, 2000, that the Federal Reserve increased its cash reserve inventory from $150 billion to $200 billion during 1999, delivering $80 billion to banks, thrifts, and credit unions during the fourth quarter of 1999 in contrast with the $23 billion distributed during the same period the year before.[21] With much less demand for cash than feared or predicted for Year 2000 computer shutdowns, as well as the cost to banks for that excess cash, the surplus was returned to the regional Federal Reserve banks during the first week of January. Amusingly, it was not the best-kept secret that billions of dollars in cash would be returning to the Fed by armored vehicles in a short period of time. Uncharacteristically, no opportunists seemed to make use of that fact.

Another interesting issue was that nearly every bank was equipped with generators, previously required by regulators in order to keep the monetary systems running during routine power outages due to seasonal weather. Thus, with generators providing both electricity and heat, the question was whether the bank would open its doors to citizens needing shelter if the municipal electrical grid went down.

I remember a business call in mid-1999 to a handsome, marbled bank in one of the cities in Boston's metropolitan area, generally then called a "hardscrabble" struggling inner city. I asked the bank CFO whether he would open his doors to the city if lives were at stake, and without hesitation he replied, "Of course." This made a deep impression on me, as we both knew such an action would not be without risk. The CFO's commitment to putting people first in the event of a grid failure was a show of humanity. I regard this as an exceptional and memorably noble gesture. I am not sure many bank managers would have that courage, then or today. A brief snapshot of humanity at its best.

To avoid financial instability, federal rules and FDIC banking regulators were not allowed to disclose the Year 2000 readiness scores of their 10,400 government-insured member banks, nor were banks allowed to disclose their own Y2K compliance scores.

Stability in the nation's banking systems was considered of paramount concern for the public's peace of mind.

My work involved assisting my bank clients with D&O liability insurance renewals during that vulnerable period. Underwriter review of Y2K readiness was crucial not only for renewal rating, but for a willingness to renew at all. Keeping bank clients from being refused insurance renewal and considered a "distressed line" was a challenge. Underwriters love actuarial tables, historic data, and trends. Underwriters hate unknowns and surprises. My habit had been to never keep anything from my underwriters, and nothing from my clients (well, except for how the sausage is made). The only thing I remember ever keeping from my underwriters was the full-grown wolf allowed to roam the cafeteria of Concord's Alcott School among children from toddlers to teens. I regret that to this day, but that's another story.*

THE CORPORATE DIRECTORS AND OFFICERS INSURANCE PROBLEM

Directors and officers liability is one of the most complex areas of liability, both for Y2K and now. It is complex in that the liability exposures for corporate directors and officers lie in disclosure and nondisclosure; cost reserving disclosures, required noncompliance disclosures, as well as corporate and individual liability. I had long been aware that most corporate officers from every sector, although accustomed to knowing the functioning of every aspect of their enterprise, were often ill-equipped to know anything about the technology division responsible for company information and computers.

As a consequence, both officers and directors were entirely dependent on their IT departments to assess all functional

* It was the idea of one of my nonprofit clients that brought animals into institutional settings. Other visitors included golden retrievers, rabbits, and toads.

capabilities, including Y2K compliance. Those disclosures have to be at the mercy of corporate technology officers, now known as CIO or CTO officers. An onerous section of the Securities Exchange Act of 1934, Section 10(b) (pre Sarbanes–Oxley Act) makes it unlawful for the sale of any security omitting a material fact. Thus, criminal penalties could be imposed for such securities law violations, exposing directors and officers if Year 2000 compliance problems were hidden from investors.

The prospect loomed that shareholders might bring suit against directors and officers for failure of due diligence if corporate management had not aggressively sought insurance proceeds to cover remediation expenses. Lawsuits might also be filed by shareholders when stock values fell as a result of corporate Y2K problems being publicized, or for inflated pre–01/01/2000 compliance statements.[22] This was clearly a damned-if-you-do, damned-if-you-don't issue, more politely described as a double-edged sword. Thelan, Marrin, Johnson & Briggs, LLP, stated, "By law, investors in your company . . . are entitled to know all 'material' facts when they decide to invest." They went on to say that "All companies—even private ones—are subject to investor fraud claims for material nondisclosures," and "In addition, corporate management has a legal duty to shareholders to act responsibly in protecting their companies."[23] Shareholder and derivative suits for compliance issues and disclosures were a formidable weight on the heads of corporate directors and officers as the twentieth century drew to a close.

As far as D&O insurance for Y2K was concerned, it is my opinion that the problem was binary and insurers were trapped. If the insurer placed Y2K exclusions on the D&O policy renewal leading up to 01/01/2000, the policy would not be of any value for the insured to renew. The insurer's D&O market would inevitably collapse. As a consequence, D&O insurers were not adding exclusionary endorsements to their policies, but were extremely diligent in determining client Y2K compliance for renewability and rating policies accordingly. A December 1998 article in *Best's*

Review reported that John F. Kearney, chief underwriting officer of Executive Risk Specialty Insurance Company, stated, "Exclusions aren't commercially acceptable today; we simply do not offer proposals for risks that we find haven't adequately addressed Y2K."[24]

In addition to D&O liability insurance matters, another issue facing directors and officers was the lack of any other form of insurance for Y2K problems at affordable prices, if at all. Directors and officers are exposed to shareholders' or damaged parties' suits for not carrying sufficient coverage. Interestingly, while many D&O liability insurance policies once had standard exclusions for "failure to carry adequate insurance," many insurers had done away with that exclusion, exposing D&O insurers to defending those suits as January 1, 2000, approached. A "failure to secure insurance" claim is far enough removed from Y2K to suggest defense would need to be tendered.

Directors and officers also faced the issue of "late notice" on their D&O policies, which has specific terms regarding the timing of notification to the insurer of any potential claim or incident that might give rise to a claim. As January 1 approached, the no-win dilemma was 1) unreported known compliance issues in earlier years could be interpreted as "late notice" by an insurer, voiding an existing policy; 2) timely notice of a possible Y2K claim to an insurer could trigger public disclosure notification obligation failure; and 3) notice of Year 2000 noncompliance satisfying current policy notification terms would surely affect policy renewal(s).

I had experience with what appeared to be corporate officers' attempt to provide advance defense for themselves for potential Y2K failure by creating "credible deniability" of known Year 2000 compliance issues. Looking back, this strategy both avoided giving insurer notice, thus avoiding jeopardized renewal terms, as well as offering evidence of due diligence by taking (unnecessary, in my opinion) termination action against some of the very people charged with successfully delivering solutions.

Evidence of preparedness was important to me in considering insurance renewals. I was ultimately willing to leave one of my largest technology clients when they refused on several occasions in early 1999 to disclose their Y2K compliance plans just prior to their insurance program renewals. It would have been both painful and a great sacrifice to let this client go, but I considered it unethical to not hold the line; continuing to work with a technology entity that had potentially not prepared for Y2K would be asking their insurer to take on far too much risk. I was relieved to finally be presented with the Board's Y2K compliance data.

An interesting quote was offered by Andrew M. Pegalis, Esq., attorney and former risk manager, in his April 1997 article directed to risk managers. Pegalis looked back at a 1932 Supreme Court decision and pointed out that:

> D&O [insurance] coverage protects corporate decision makers from suits brought against them personally for negligent business decisions. Even if an entire industry neglects to take pre-emptive measures, directors and officers could still be held liable. As Justice Learned Hand eloquently stated in the 1932 T. J. Hooper opinion, "There are precautions so imperative that even their universal disregard will not excuse their omissions." As evidence of negligence, shareholder suits will likely cite material reporting failures. Under the SEC's S-K regulations, directors must report exposures to the millennium bug and compliance to the SEC.[25]

Pegalis's original feature article in *Business Insurance* in December 1996 took a very uncompromising position with vendor contracts; "[Y]ou must demand unusually stringent guarantees" and expert legal advice.[26] He suggested web sources for contract wording, as well as for pointing out specific accounting problems with cost allocation(s).

J&H Marsh & McLennan noted in its client Y2K publication that failure of corporate management to make required disclosures could possibly trigger policy exclusions for fraudulent, dishonest, or criminal acts.[27] In a mid-1995 article in the *Wall Street Journal* Europe edition, UK Withers law firm partner stated: "Absolutely, directors and officers won't be liable for honest mistakes, but they can certainly be held responsible for gross negligence. It's getting far too late to claim ignorance about millennium issues."[28]

Opinions varied widely on this matter. A mid-June 1998 article from London reported, "The Insurance Information Institute in New York said most U.S. insurers are not using exclusions, saying that they are not needed because policies do not cover Year 2000 problems anyway." However, the article's author, Patricia Vowinkel, went on to conclude that with so much uncertainty on both policy language and exclusions, "the only winners are likely to be the lawyers."[29]

With specific and lengthy Security and Exchange Commission directives on Year 2000 disclosers for publicly held companies during the 1990s, Ernst & Young announced in June 1998 that only 29 percent of insurers were willing to disclose their underwriting exposures although 62 percent of those insurers had "made an effort to calculate their costs." Ernst & Young Director David Holman, speaking before the Insurance Accounting & Systems Association (IASA), stated that 86 percent of respondents to an informal survey "needed to get rid of 'boiler plate' disclosure language currently in use in their public documents." Holman added that the Y2K fortuity issue should help to protect insurers from responding to Year 2000 claims. However, there were exceptions: directors and officers, errors and omissions, workers compensation, and medical malpractice. He also added automobile claims in the event traffic lights should fail. "The industry is not out of the woods yet," Holman concluded.[30]

Earlier in the year, in a much more optimistic vein than Ernst & Young's survey, the *CPCU Journal* reported that Sean Mooney, senior vice president and chief economist for the Insurance Information Institute, stated that "most large and mid-sized insurers are ahead of the curve in fixing their millennium bugs."[31]

Nevertheless, there were other cases where accountants, in addition to attorneys, were comfortable pointing out vulnerabilities of the insurance industry side without mentioning their own exposures to Y2K claims, litigation, or compliance.

TECHNOLOGY INVESTMENT

Regarding investment houses, a brief mention that securities and brokerage houses distributed glossy publications on the Year 2000 computer issue for their clients. Both the Gartner Group and Cap Gemini were cited regarding spending estimates and costs. In Salomon Smith Barney's mid-1999 publication, Cap Gemini's estimate of $850 billion worldwide expenditure noted; "We believe the Y2K spending will likely be the single most significant event in the Information Technology services industry." Smith Barney went on to reassure clients that doomsayers were largely opportunists; "So-called 'Y2K Gurus' have created a substantial business for themselves, calling for massive power outages, rioting, starvation and disease, because of the Millennium Bug." Summaries of the readiness of the major economic sectors, government, and litigation notes accompanied a list of five technology stocks to own (IBM, Microsoft, Cisco Systems, AOL, MCI WorldCom). Salomon Smith Barney's own Y2K compliance work followed; its "Conclusion" was that fear and panic, (F.U.D.—fear, uncertainty, and doubt) were the largest factors in the new millennium's uncertainty.[32] Reassuring investors was their primary mission—a well-crafted narrative.

On the investor's side, technology spending was analyzed with respect to Year 2000 compliance pressures with an estimate of 20 to 40 percent of technology budgets for Y2K. The Gartner Group of Stamford, Connecticut, surveyed 65,000 companies that reported compliance costs would delay technology spending on many new applications. Interestingly, Giga Information Group stated that the only spending not being postponed was e-commerce. While revenue growth for desktop computer manufacturers like Dell and Compaq increased, enterprise software vendors like Peoplesoft and J. D. Edwards decreased. Added to modest technology growth predictions was an acknowledged lack of available technology talent.[33]

The Gartner Group was reported in the *New York Times* to have been consulting clients on Year 2000 issues since the late 1980s, beginning its broader analysis of the problem in 1993. Gartner's Y2K analysts multiplied their estimated lines of computer code by the cost to hire outside computer programmers to come up with the startling figure of $300 to $600 billion (a remarkably large delta variable). Critics said Gartner's research process left many unanswered questions. The outside figure of $600 billion was disputed by analysts, including International Data Corporation and Morgan Stanley Dean Whitter, but ultimately became the worldwide estimate for remediation instead of the original U.S. estimate. Gartner's prominence in the Y2K field was never shaken, however. Competitors Cap Gemini, Giga Information, Jupiter, and Forester acknowledged Gartner's singular leadership, based in some respects on their global range and widely distributed survey results. Deutsche Bank's Edward Yardini offered, "Gartner's methodology leaves something to be desired, but it's as good as it gets."[34] Yardini was widely quoted for his analysis of corporate readiness, surveying most of Standard & Poor's stock index companies and reporting troubling statistics on readiness.

Investor confidence in Y2K technology services stocks dropped sharply during 1998, according to the Bloomberg Year 2000 Index as reported in the *New York Times* in late October 1999.[35] What were very hot Y2K software companies in 1997 and early 1998 fell from +60 percent to -20 percent as investors worried about post–Year 2000 operational survival. An earlier *New York Times* article of mid-July looked at those Y2K-related software service stocks—Information Architects (formerly Alydaar Software Corporation), Keane, Peritus Software Services, and Viasoft. Those names were very familiar big stars in the computer compliance field for Year 2000. In mid-1999, Viasoft's stock value went from a high of $65.25 in 1997 to $6.3438; Peritus from a 1997 IPO of over $30 to 21 cents; Keane from nearly $60 in 1998 to $22.9375; Information Architects from nearly $30 in mid-1997 to $2.125.[36]

Clearly, redefining, renaming, and marketing broader technology services was a struggle for the more prominent Year 2000 software services; their stock prices reflecting overly dramatic slides by 1999. Viasoft, for example, was reportedly struggling to position itself for services such as Euro currency conversion, e-commerce, mainframe legacy systems management, client-server networks, internet applications linking, and most challenging, moving from a product to a service-oriented organization.[37]

Year 2000 not only posed challenges for software compliance in every sector, but also it caused disruption to the normal flow of business opportunity. Whether or not there were any valuable operational lessons learned is difficult to assess. Very little about Y2K compliance was mentioned following the millennium turn. We can only speculate on the reasons. First, corporate management has always been reluctant to advertise their failures. Next, 9/11 occurred less than two years later, and all national focus shifted to domestic terrorism. Third, for the many entities using the ten-year "windowing" approach to the January 1, 2000, deadline, Y2K errors finally arising in 2010 could be reported as no more than a "system problem" with scant detail. Again, this is a contemporaneous story of the 1990s about the actions of people who did not know the outcome of their own, their country's, or the world's Y2K compliance efforts.

4

YEAR 2000 IS NOT COVERED

Y2K and Insurance

This is my December 1998 letter to my clients regarding Year 2000 and their insurance.

N. P. JAMES INSURANCE AGENCY
33 BEDFORD STREET
CONCORD, MA 01742
TELEPHONE (978) 369-2771 FAX (978) 369-2778
npjames@compuserve.com www.npjames.com

YEAR 2000 Statement

December 21, 1998
Client
Company
Street Address
City, State Zip Code

Dear *Client:*

Since mid-1997 I have been including Year 2000 Alert notices in renewal advisory letters to all my commercial clients. I have been alerting everyone to Year 2000 issues and am writing to you as a new client because it is vital that you understand the issues involved with Year 2000 and your insurance coverage.

First, may I say that I am not a Year 2000 technical expert, nor can I or anyone else at this time assure you of what will and will not be defended and/or covered with respect to Year 2000 losses or claims. My remarks around the country and in London this past July involve the range and magnitude of issues and influences surrounding Year 2000, with some predictions as to trending [for a copy of the text of my remarks, please call me]. The issues affecting coverage as I see it today:

A Question of Coverage and Exclusions to Coverage Insurers never contemplated nor now contemplate covering Year 2000 technical problems under the standard insurance contract wording (including but not limited to issues of fortuity, latent defect, programming error, etc.). Both the standard policy exclusions as well as new Year 2000 Exclusions serve to limit/eliminate coverage further.

Insurer Insolvency Estimates of the cost of Year 2000 fixes have been $600 billion, with the bar estimating litigation costs reaching $1–$1½ trillion. However, the total U.S. insurance reserves are but $280 billion, thus, a total collapse of the insurance market could ensue from wholesale claims demands and defense costs. This means not only would you be unable to collect on your claim and your defense costs would go uncovered, but even the mortgage on your personal home would be called in due to your inability to get the mandatory insurance protection required by your bank. It is unrealistic to expect that commercial insureds can anticipate collecting any more than pennies on the dollar if insurance is asked to cover the preponderance of costs to remediate Year 2000 technical problems.

Year 2000 Expense Measurement Defenses Some Year 2000 technical experts have calculated, by looking at legacy systems and the past costs of two- versus four-digit date fields as well as the use value of money over time, that Year 2000 remediation costs to be far less than the additional memory space in past technology. This makes a reasonable defense: "no monetary loss, therefore no claim!"

Government Intervention My mid-1997 remarks to the American Bar Association Annual Meeting suggested that Congress could *declare* Year 2000 a "virus" or more likely, a "natural disaster," both possibly invoking coverage which we would not contemplate under current insurance forms without such declaration. What both Congress and state legislatures are doing first, interestingly, is enacting legislation holding themselves harmless from their own Year 2000 processing failures. [My first foray into government predictions appeared in a May/June 1997

Boston Bar Journal editorial co-authored with Boston attorney Barbara O'Donnell, where I suggested that the IRS, even if its own systems were hopelessly snarled in Year 2000 problems, would, nevertheless, be very unforgiving of late tax filings, imposing late filing penalties.] It is my opinion that, given a conspicuous lessening of liability by the public sector, there will be considerable conflict with the private sector, which appears to be held absolutely liable.

Year 2000 Information and Readiness Disclosure Act In October, 1998, President Clinton signed into law the Act, which provides certain limitations to liability in civil actions. Your attorney can advise you on how this act protects you, and I suggest that you call your counsel.

Millennium Mediation Council In an editorial for the *Insurance Times*, I have suggested a Millennium Mediation Council, to mediate claims before 1/1/2000. In my opinion, it is a solution for directing all resources toward fixing the problem, and not to litigation and court costs [I am including a reprint of my article.] Reviewing www.itaa.org shows trending toward a limitation of liabilities for Year 2000 problems.

Specific Year 2000 Insurance To date, outside of a few Directors & Officers insurance forms with specific Year 2000 coverage, there is no affordable Year 2000 coverage on the U.S. market today. If you directly request it of me, I will keep you posted as to any emerging markets.

Conclusions Now, it is important that I state that I have no control over the insurance market other than participating in discussions as outlined above. Insurers know their possible vulnerability, and may have already asked you to complete a questionnaire. I have been advised that such questionnaires from all parties are not privileged and are discoverable, thus, great care in their completion is imperative. Of course, I am not and cannot advise you that Year 2000 problems will be either defended or covered by your insurer for the reasons stated above. I can tell you that insurers, the courts, and our state and federal governments will be determining coverage as Year 2000 approaches.

What Are Our Roles? As I have been advising you, your current insurer will probably attach a Year 2000 Exclusion to your policy for exposures we have identified. Under those circumstances, should you seek a new insurer? No. Historically insurers look to protect their current clients to the greatest level possible. Applications to a new insurer with Year 2000 exposures are being declined. However, I will take whatever necessary steps you instruct me to take to seek alternative coverage. It is my goal to work with you to make very clear to your insurer your

Year 2000 exposures, and to make very clear to you whatever Year 2000 Exclusions your insurer wishes to attach to your renewal policy. We will work together with you to keep communication lines open in order to strengthen your relationship with your insurer.

What I Ask You to Do: Please show this letter to counsel, and advise me of your attorney's response. We can work together to protect you as best we are able. I will be happy to meet with both you and your attorney to discuss your insurer's position and intentions on Year 2000 coverage and lack thereof. It is my hope that the estimates of massive failures will be diminished as 1999 progresses, and that many of these matters will be moot.

Finally, just FYI: The Concord Chamber of Commerce has asked me to speak at and moderate a Year 2000 panel January 22, 1999, at Newbury Court. Speaking mostly to public safety issues as we face infrastructure problems in the northern hemisphere in January, 2000, was Concord's Town Manager, Police Chief, and Light Plant head, as well as Representative Pamela Resor and Congressman Marty Meehan's Year 2000 specialist, speaking on legislative action. The Concord Chamber will be hosting an update series, probably in the fall. If you are interested in attending please ask the Chamber to notify you.

I will expect to hear from you soon.

Sincerely,

Nancy P. James

Nancy P. James used this rubber stamp on all insurance policies mailed in 1999. *Courtesy of the author.*

It's possible that all but insurance wonks will want to skip this chapter. Much has been chronicled from the technology sector about Y2K remediation and compliance efforts and solutions. Less is written about what is regarded as the back-office insurance sector. This chapter is intended to describe the issues facing insurers, particularly during the last decade of the twentieth century, the runup to the new millennium. As January 1, 2000, drew closer, self-preservation forces, competition, liability shifts, and disclaimers heated up essentially across all commercial sectors. The contemporaneous documents of those pressures tell a story about those who did not know the outcome of the Y2K issue, primarily as related to liability, risk, and insurance.

My agency letter to all of my business clients dated December 21, 1998, tells the entire story as of that date regarding insurance coverage and claims issues. In that letter I addressed the stakes as plainly as possible: at that point, estimates of the cost of Year 2000 fixes were $600 billion, with the bar estimating litigation costs reaching $1–$1.5 trillion. However at the time, the total U.S. insurance reserves were estimated at under $300 billion. As a consequence, Y2K claims demands and defense costs would result in a total collapse of the insurance market.

That outcome would be catastrophic to insurers, the hundreds of thousands of insurance industry workers, the American economy and, of course, every American. Businesses, including banks, would have no insurance protection and would likely collapse, many due to lack of the necessary regulatory insurance protection. The personal home and auto claims of Americans would go unpaid, all bank loans to consumers would be called in for lack of required insurance, and no personal property would be protected. And surprisingly to some, taxpayers would pick up all of those losses as a consequence of state statutes enacted for insurer insolvencies.

Thankfully, none of that came to pass due to the diligence of Year 2000 remediation. But insurers were understandably very nervous heading into 2000 since they knew the consequences if Y2K costs were to be borne by insurers.

At that time I was also working on an advisory basis with a major insurer to review their insurance policy forms and offer my opinion on whether or not Y2K might be covered under any of their policy terms.

Insurers were marshalling resources to determine possible coverage triggers and plan defenses against baseless claims assertions. In fact, my own insurer, Utica National Insurance Company, one of the largest insurance agency errors and omissions liability insurers, refused to renew my coverage leading up to January 1, 2000—because my clients were primarily technology-based enterprises. Too big a risk, they determined, although I had been a Utica insured for over 15 years with no claims.*

By 1999, a collaborative Insurers' Year 2000 Roundtable had been established with thirty-three property/casualty insurers and reinsurers to address coverage issues.[1] State and federal legislative and regulatory bodies were dictating insurers' entitlement to claims defenses, described in chapter 6. Attorneys for both plaintiffs and insurers were assembling teams for lucrative battles.

You may still want to skip over this deep dive into Y2K and insurance! I would note that for those interested, the main points are summarized at the end of this chapter.

NOW ADDRESSING WONKS (THIS IS THE INTERESTING PART OF MY STORY)

Reporter Penny Williams ran a front-page story, "Millennium Liability Slated to Hit Insurers, Agents, Lawyers and More," about my Year 2000 work in the *Insurance Times* on March 3, 1998. I was quoted as saying, "The [Y2K] liability is only limited by the limits of one's imagination."[2]

Williams quoted my reference to an estimated $600 billion cost for remediation, and the fractional total insurance industry capacity. I predicted, "Insurance surprisingly will most probably not be a viable place to turn for Year 2000 relief," projecting that insurance carriers were poised to protect themselves with Year 2000

* Fireman's Fund, which was a newly emerging insurer for insurance agencies, picked up my coverage for the next several years until they left the market, at which time Utica again underwrote my coverage.

exclusions or severe sublimit of liability coverage that would be imposed on all commercial renewals including D&O and E&O liabilities. The article continued:

> "Insurance forms drafted decades ago have no exclusions for Year 2000 disasters . . . current insurance policies would conceivably be called to respond to both first-party and third-party Year 2000 claims. Or, at the very least, have a duty to defend imposed. Courts may rewrite insurance contracts to find coverage," James said. She also predicts that the federal government may step in to *help* the industry by possibly declaring Year 2000 a "virus" or a "natural disaster," either of which could affect the issue of fortuity, and invoke coverage not anticipated at this time.[3]

Invoking coverage by fiat was not entirely improbable; it was clear I was becoming uneasy.

Williams also interviewed Leslie Fellows, Insurance Services Office (ISO) spokesperson, who referenced a proposal for two general liability endorsements; one excluding any injury or damage for Y2K failure(s), the other allowing insurers to underwrite and cover specific operations where Y2K compliance remediation had been made and tested. Fellows added that an ISO property form exclusion would be drafted as well as a business interruption exclusion form with a coverage grant of up to $25,000 for Y2K related losses.

Commenting on Fellows's statement, I asserted: "'ISO maintains that since Year 2000 risks have not been contemplated for coverage, the exclusion forms are more a Coverage Clarification Endorsement. . . . It will be difficult to impossible to negotiate these limitations away.'"[4]

I was also quoted in a September 1997 *ComputerWorld* article, "Insurers Plan Limitations on Y2K Coverage": "Customers who assume that existing policies provide Year 2000 coverage would be well advised to check again. Insurance companies don't want to have to pay the bill for damages that some observers estimate could reach hundreds of billions of dollars."[5]

I will pause here to look back at similar public and private pressures on the insurance industry that informed my opinion. In the 1980s, for instance, a New Jersey court refused to honor the pollution exclusion on a municipality's insurance contract, forcing the insurer to finance the toxic waste cleanup of the municipal landfill. Subsequent courts following suit caused the municipal insurance crisis facing the entire nation in the mid-1980s.

President George W. Bush's first words following the attack on the World Trade Center declared 9/11 an "act of war." It was a long 24 hours later—after likely being advised that war was universally excluded on all standard business insurance policies—that President Bush declared the attack an "act of terrorism." Insurance would then apply for the billions in 9/11 life and property losses. The language used to describe such an event makes a staggeringly significant difference.

Public pressure and sentiment can have a similar impact. When Huirricane Sandy struck in 2012, southern attorneys were widely publicized for demanding that insurers cover the massive flood losses, even though flood was a peril clearly excluded on most policies. Then came the COVID-19 pandemic, where public pressure was immediately exerted to try to force insurers to cover business losses from shutdowns, ignoring the "communicable disease" exclusions in most business policies. Those same pressures were exerted on insurers to cover Year 2000 software remediation costs, which had never been contemplated nor rated by underwriters.

In the technology, insurance, and law-related documents during the 1990s run-up to Year 2000, there is exhaustive, arcane detail. An extensive chronicle of the content of those documents is more appropriate for a graduate thesis, not for my odyssey during that time. Here I will outline the dilemmas and pressures facing insurance professionals, insurance carriers, insureds, agents, and advisors.

Insurance issues were crucial from the very beginning, since insurance proceeds arguably finance the costs of defense and litigation where coverage applies. In addition to investing millions of dollars to update its own in-house technology, the insurance industry had to brace for an explosion of lawsuits surrounding the failure of computer systems to be Year 2000 compatible. Of course, the applicability of coverage for Year 2000 remediation costs was the issue. A trillion-dollar issue! Thus, worthy of debate. By mid-1999, insurance professionals were getting cranky with cost projections from "experts," arguing that most estimates were pure speculation.

Generally, the majority of Year 2000 insurance issues involved commercial policies—those insurance contracts for businesses, mercantile, manufacturing, and corporate management. It was suggested at the time that Y2K disruptions involving homes and automobiles would ultimately end up as claims against manufacturers' liability.[6]

John Liner Letter, one of the insurance industry's most respected analysis publications, offered its prediction in May 1998 of likely litigation: "breach of contract, defective products, infringement of intellectual property, and directors' and officers' liability." Commercial general liability policies were not expected to cover Y2K, since such losses are "considered business risks."[7]

Y2K was a classic example of disputes over coverage: assertion of coverage under existing policy forms, the tangibility of software, policy exclusions, definitions of Y2K failure causes, wisdom of the introduction of Y2K exclusionary endorsements, and more. Defining Year 2000 problems as a "virus" or using "force majeure" protections for contracting parties were posed for the purposes of finding insurance funding (in the case of "virus") or legal protections ("force majeure") in Y2K noncompliant contract disputes. *Best's Review*'s editor Caroline Saucer quoted a J&H Marsh & McLennan executive in a May 1998 editorial questioning exactly when Y2K coverage would be agreed to be in effect (this is called a "coverage trigger"): policies in effect when the software code was written, when the product was installed, when the defect should have been discovered, or when the defect was discovered?[8]

A year earlier, *Best's Review* also ran an article by editor Saucer, theorizing that an argument could be made that "property damage to the insured's computer system has already occurred, and the conversion costs are a covered loss."[9]

A month earlier, in April 1997, the "Risk Strategist Web Site—RiskFocus Item" offered ten steps to assessing Y2K risk as well as "Finding Risk Solutions in the Marketplace." The site suggested there were "three sources of potential coverage to be explored." These included third-party liability, D&O liability, and property/business interruption. The fortuity issue (the foreseeability of Y2K disruptions) was raised, with advice to readers to "work with your broker well in advance" and "subject to the policy's terms and conditions," offering only the likelihood of ensuing fire being covered.[10]

The *IIAA* (Independent Insurance Agents Association) magazine in its December 1997 issue had a feature article on Year 2000 where the author declared, "Anxious to protect itself against catastrophic Year 2000 payouts—and to make a buck on the problem where it can—the insurance industry has begun to crank out a variety of weapons in the war on the so-called Millennium Bug."[11] Additionally, in early 1998, IIAA Technical Affairs sent an eight-page letter to association members very critical of ISO's Y2K exclusionary endorsements, seemingly arguing in support of Year 2000 remediation coverage on existing policies.[12] IIAA's letter argued incorrectly in my opinion on the matter of data tangibility. In both cases, IIAA's positions struck me as odd from an insurance agent's association, knowing the magnitude of the issue and its catastrophic consequences to the insurance industry.

In any event, it was clear that anxiety was growing and defenses were being built.

The issue of data and software being "tangible property" arose again as Year 2000 exposures were examined. With regard to the tangibility of data issue, I authored an article with the assistance of attorney Ann R. Truett, published in 1993.[13] The important *Magnetic Data* case of 1984 ultimately left the tangibility matter undecided in 1989, thus leaving the matter open until the 1991 *Retail Systems* case determined data was tangible based upon recent tax decisions. At that time court cases were few and varied in their findings, so Truett and I looked elsewhere, finding that the Uniform Commercial Code had declared software a "good" for taxing purposes in every state, and it appeared at the time data would ultimately be declared "tangible."

From my perspective, the issue of data tangibility for insurers was rendered moot with the subsequent inclusion of a specific definition of data and software as "intangible" within the insurance policies property definitions. However, the issue of tangibility was anything but moot as the new millennium approached. The state of California has been particularly contrary to East Coast courts' coverage interpretation over the decades, as demonstrated by an impressive volume published by the California law firm Cooper, White & Cooper, LLP. The firm concluded that the matter was anything but resolved and would be a major factor in Year 2000 litigation.

Cooper's examination of various Y2K issues of contention was offered, section by section. "Occurrence," "fortuity," "duty to defend," "insured's intent to inflict harm," "breach of contract" awards, as well as broad interpretations of "damages," offered a sweeping assertion of coverage triggers for Y2K remediation and compliance. In the matter of "occurrence," intentionality is considered ambiguous with no consensus among courts. Robert Sallander of Cooper, White & Cooper maintained that an insured's intention to use noncompliant computers makes no difference, but the insured's *subjective* intent to actually cause property damage is the issue. It is not an *objective* reasonable standard of the insured's intent to inflict harm, Sallander argued, but the "insured's subjective state of mind that must be shown." Triggers of coverage were outlined; injury-in-fact, exposure, manifestation, and continuous trigger, with law firms hoping to invoke multiple insurance policies with a continuous-trigger argument.[14]

In stark contrast to a very liberal California interpretation of "fortuity" and "tangibility," English law was expected to respond differently. An *Asian Insurance Review* article in early 1998 by Michael Graham, partner at Barlow Lyde & Gilbert of London and Hong Kong, offered that under English common law, losses not fortuitous are simply not covered, with Y2K fortuity being weak. Graham added that English courts had not determined that software is "property" (tangible), and would likely be a matter of dispute for Y2K compliance software issues. He anticipated that only subsequent Y2K-related damage such as theft or fire would likely be covered. Those differences aside, Graham concurred that the global cost of Year 2000 compliance would be US$1 trillion with damages and punitive awards approximating US$100 billion.[15]

Taking yet a different position, Boston's *The Standard* weekly magazine cited two studies, one by Milliman & Robertson, the other by Standard & Poor in mid-July 1999, estimating a maximum of $35 billion in claims and defenses costs to the U.S. insurance sector. The prestigious American Insurance Association (AIA) disagreed publicly, calling the study conclusions "extremely speculative;" that factors not considered "could contribute to an assumption of insurance obligation for Year 2000 disruptions that have no basis in fact, contract or law." Milliman & Robertson defended their estimates, arguing that the insurance industry's costs would not be minimal with the likelihood that parties would not be absorbing Y2K costs without attempting insurance policy coverage response. S&P, reluctant to put a figure on costs, did estimate the costs as significant, identifying vulnerable areas of risk like directors and officers, third-party professional liability, as well as business interruption expenses.[16]

Many coverage disputes were being anticipated on both the insurance and legal sides. One interesting dispute: "All risk" coverage implies coverage unless excluded; Y2K, if defined as a "peril," would then argue for coverage to apply. In a "named peril" policy, again, if Y2K is a "peril" and not listed as a "covered peril," then coverage would not apply. This position, however, had considerable weakness, since arguably any new threat might be given a brand-new designation, or name, to invoke coverage. Nevertheless, it was long established that general liability insurers, accepting the primary risks of bodily injury and property damage, were not intended to be the guarantors or warrantors of their insured's workmanship or product, nor for the withdrawal of any third party's defective non-Y2K compliant parts.

Directors and officers liability topped the list of all exposures in my mind, with professional and general liability following, then surety. For professional and general liability insurance specific to Y2K exposures, contract breaches would trigger claims for non-compliant products and services failing definitive attachment of Y2K exclusions. Liability for failed heating systems causing frozen pipes, liability for failed security systems, embedded noncompliant chips causing automobile injury or death; each has a probability for liability claims. Surety policy claims could be presented for contracts unable to be completed on time or make payments. Medical malpractice liability is exposed if records became inaccessible or medical services unsuitable due to Year 2000 issues. Any professional service could experience malpractice complaints should their services be interrupted.

Property policies normally cover fire losses regardless of the "proximate cause." Eventually pre–January 1, 2000 quarrels over coverage were more focused on Y2K *liability* matters and *business interruption* claims than on property-related claims.

Business income losses, both primary (i.e., shutdowns from a closure following a fire) and contingent (i.e., supply chain interruptions) come next on my priority list, with property and loss-of-use losses coming last (last, because property loss values can almost always be calculated and known in advance). Year 2000 coverage exclusions would certainly reduce claims, but an insurer's duty to defend would need to be determined by the courts when a claim was presented.[17]

Year 2000 exclusions were prohibited for Workers Compensation policies. As a matter of public policy, all workers injuries need to be covered by state-mandated Workers Compensation coverage benefits. Workers Comp benefits, specified on a state-by-state basis, cover employees for any work-related injury, which could be significant in the event of power failure, machinery breakdown, or other hazardous work issues affected by Y2K failures.[18]

The European Union's complaint that the United States was slow to adopt exclusions was reasonable. *The Standard*, a well-respected New England insurance weekly, published in mid-July 1998 that the IIAA had just withdrawn its opposition to Year 2000 policy exclusion endorsements.[19] That was one week before I spoke in London, a time when most of the European nations had already been adding Y2K coverage exclusions to their insurance policies. IIAA was concerned about overzealous use of exclusions, according to *The Standard*, but felt that both regulators' affirmative action as well as underwriting by insurers would prevent abuse.[20] I remain as bewildered now as I did at the time to a dangerous lack of understanding of the potential Year 2000 claims issues on the part of agents and their associations. Or was IIAA just posturing for the sake of siding with insureds? In a May 1998 article in *The Standard*, representatives from Aon Risk Services and Willis Corroon Americas, speaking in a national satellite broadcast to the CPCU Society, stated their frustration with insurance companies not being definitive on policy coverage, with Eric Ball, senior vice president of Willis, stating that "there could be more problems for insurers that do not indicate how they stand, and then try to exclude everything."[21]

I have notes from a conversation with an ISO spokesperson on August 6, 1996, regarding states accepting Y2K exclusions. At that time, forty-two states had accepted the exclusions, with Alaska, Louisiana, Massachusetts, Maine, Missouri, Nebraska, Texas, Washington, DC, and Puerto Rico not approving. The spokesperson said the ISO received at least ten calls a day from insurance companies asking when exclusions will be approved, noting that

Massachusetts had promised in July an answer "within a month." The spokesperson went on to say that their sense was that even when approved, insurers would wait to attach Y2K exclusions to policies. This, indeed, turned out to be the case.

Two years later, on August 11, 1998, forty-seven states had accepted Y2K insurance policy exclusions, leaving just Maine, Massachusetts, and Alaska. Quoted via NewsEdge, ISO spokesman David Dasgupta stated that exclusions were necessary because insurers had not collected premiums for Y2K losses. ISO, representing 2,900 insurance companies and primarily responsible for most forms used by insurers, drafted the widely approved and accepted Y2K exclusion forms. While NewsEdge reported insurance brokerage firms declining to comment, law firms were vocal and critical of the exclusions.[22]

Dasgupta's premium rationale was not widely held, since there were no previous data on which to calculate exposure and premium rates within existing policy forms. More importantly, insurers were taking the position that Year 2000 costs and losses did not meet many terms of existing, long-used insurance contracts. Coverage was widely regarded as a completely new and independent unit to be underwritten looking forward and not back, with specific Y2K-related criteria for coverage.

A July 1998 article in *The Standard* reported on a meeting of the National Association of Insurance Commissioners (NAIC), in which Massachusetts Director of the Division's State Rating Bureau Stephen J. D'Amato was vocal in his concerns regarding Y2K exclusions. Division spokesman Michael Klein, explaining why the commonwealth was reluctant to approve the exclusions, added that regulators were unclear on what insurers were trying to do with them. Adding a very odd comment, Klein expressed concerns that "policyholders were not consulted in the formation of the exclusions." With Massachusetts leading the opposition to Y2K exclusions, D'Amato and Klein used the "lazy underwriter argument," asserting that, with blanket use of exclusions, underwriting Year 2000 exposures would be unnecessary.[23] Klein's objection was that underwriters (those insurance professionals assessing risk and premium rating) were using wholesale policy exclusions instead of hands-on Year 2000 risk assessment and pricing.

A note of explanation for Massachusetts' delay in accepting the exclusions: The Massachusetts Commissioner of Insurance is traditionally a political appointment in the commonwealth, thus one fraught with controversy and often conflict. Linda L. Ruthardt was Governor William F. Weld's appointment in 1993 and served for an unusually long nine-year tenure. Excerpting the *Boston Globe*'s 2010 obituary: "Ms. Ruthardt's tenure, ending in 2002, was unusually long for the lightning rod post of top insurance regulator. She was sometimes called the good news–bad news insurance commissioner. 'Whoever is commissioner of insurance is automatically the most hated person in the Commonwealth,' Ms. Ruthardt told the *Globe* in 1996."[24]

In my opinion, the Massachusetts Commissioner of Insurance's position on Year 2000 exclusions was both foolhardy and uninformed, displaying ignorance of the staggering magnitude of the Y2K liability issue. But the usual finger-pointing was robust. In what I will at the outset call posturing, some of the most prominent national insurance risk management houses disparaged insurer's use of exclusionary endorsements, labeling insurance companies as being "paralyzed with fear." Willis Corroon's study harshly described some companies' Y2K endorsements as "indecipherable lawyerese gobbledygook," adding (as mentioned above) that those insurers who "remained conspicuously silent" presented the most danger to policyholders. A counselor with the National Association of Independent Insurers disputed Corroon's claim of insurers being unprepared for the millennium turn, offering that while some insurers are more outspoken than others, "nobody's running scared."[25]

Scott Seaman and Eileen King Bower in the July 7, 1999, issue of *Mealey's Litigation Report*, noted that the ISO computer-related commercial policy exclusions had been accepted by all fifty states by October 1998. Some states had exceptions for specific commercial policy lines and/or personal lines of coverage.[26] Insurers had been meeting during this time to discuss the uncertainties surrounding Year 2000 exposures and forming plans for exclusionary wording, underwriting, pricing, and defenses. The fear with these singularly concentrated meetings involves antitrust accusations. Again, the magnitude of Y2K uncertainties naturally focused on insurers' consistent responses.

The plaintiffs' bar presented the common argument that any new policy exclusion was evidence that coverage was previously in place on the policy. Exclusionary wording is drafted in two distinct ways: the first to limit recovery by noting a dollar sublimit of coverage, the other to specifically provide additional clarification and definition of existing policy terms. The first approach of sub-limiting recovery is by far the better course, and more successfully relieves the courts of expensive litigation for lawyers quarreling over policy term interpretation.

Y2K was a notable exception for sub-limiting due to the magnitude of potential claims and losses. Ultimately plaintiffs would need to prove coverage existed under the policy terms exclusive of coverage clarification endorsements. Any time a new exclusion or "clarifying term" is added to an insurance policy contract, it can be argued that the prior policy had a priori granted coverage, reasoning that if coverage were not inherent in the prior policy, what was the need for a new term or exclusion? That argument is logical and has consistently been used successfully by plaintiffs' attorneys. Thus, the reluctance by insurers to add Y2K exclusions without careful deliberation of the consequences. I explain in this book the reasons why Year 2000 compliance costs or remedies do not meet any policy terms for coverage. The policy forms discussed here are standard commercial property and liability forms, not professional errors and omissions forms, which are designed to protect against the financial losses associated with malfunction, failure, and/or product recall.

Opinions varied widely on this matter. A mid-June 1998 article from London reported: "The Insurance Information Institute in

New York said most U.S. insurers are not using exclusions, saying that they are not needed because policies do not cover Year 2000 problems anyway." The article's author, Patricia Vowinkel, went on to conclude, however, that with so much uncertainty on both policy language and exclusions, "the only winners are likely to be the lawyers."[27] Ultimately pressure was applied by reinsurers who declared that reinsurance would not be underwritten for insurance carriers failing to issue Y2K exclusions to insurance policy contracts.[28]

Original equipment manufacturer (OEM) products containing software and parts from a variety of sources further complicated matters for Y2K failure issues. OEM products with those same upstream software and parts issues prior to Year 2000 compliance had another complex layer of potential failure. Those issues were related more to licensing and patents, as opposed to quality control and product failure. Y2K was an added issue to deal with.

In a December 1997 insurance industry publication, a survey of Cigna, CNA, Chubb, The Hartford, Ohio Casualty, The St. Paul, Travelers, and USF&G, responses to Year 2000 coverage questions were either not answered or reported to be in study. St. Paul's executive vice president, Michael Conroy, stated that most insurers were looking to reinsurers for guidance.[29] St. Paul mailed a letter and brochure to agents that same month, pointing out the Year 2000 computer issue, including software, vendor, partner, and other systems' advice summaries and website references.[30] No Y2K insurance coverage availability was mentioned in that mailing.

Insurer USF&G Corporation in December 1996 sent the same type of letter to its agents, listing three resources following a brief explanation of the issue. Referencing in bullet form ITAA's nine-point plan, USF&G also made no mention of insurance coverage for Y2K costs or litigation defense.[31]

In contrast, in an earlier April 1997 article in the *Insurance Times*, Chubb and Son, Inc., was reported to have started working on the millennium coverage problem in 1995, devoting an entire building facility to the issue. Thelma Rosenthal, Chubb's Year 2000 senior design coordinator and project manager, said that "this early planning recently allowed Chubb to approve its first three-year insurance policy expiring in 2000."[32]

And I will note again that many insurers directed insureds to ask their *agents* about coverage!

The National Association of Professional Insurance Agents (PIA) published the most rational comprehensive summary of Year 2000 issues in 1998. In the book, PIA admitted that the question of insurance coverage for Y2K issues was undecided: "It is clear to PIA that there is a wide difference of opinion among carriers as to whether Y2K losses are covered in whole or in part under current insurance policies." Very useful arguments for and against policy interpretation, as well as coverage application, followed. PIA disputed the position that the commercial general liability (CGL) policy was an "all-perils" policy as well as disputing the common analogy of Y2K with Pollution, which would invoke Y2K coverage. An unusual argument for the requirement for equipment maintenance was included as a possible defense against coverage as well.[33]

It is important to pause here to point out that insurance risk management advisory firms to our nation's largest commercial enterprises were in a particularly vulnerable position during the late 1990s runup to 2000, since so much about Y2K was both unknown and unknowable. Those houses were singularly relied upon, reasonably or unreasonably, by their clients to make definitive statements of coverage, risk, and liability.

My own letter to my clients was a full two and half pages explaining why insurance coverage for Y2K problems would be unlikely (my letter in its entirely is included at the end of this chapter). Other agencies were a bit more optimistic and, in my opinion, a bit misleading. A Sullivan Insurance Group letter dated July 1999, states: "Insurance buyers can not [*sic*] rely on insurance coverage for failure arising from non-fortuitous activities emanating from things such as systems designs, computer code or the failure to remediate non-compliant systems." The letter also states: "Year 2000 policies are available for many businesses from certain insurance carriers; please call us if you would like to discuss this coverage." The concluding paragraph offered: "Any of our staff are prepared to discuss your concerns and answer any questions you may have about your specific insurance coverages."[34] While I had the greatest respect for the Sullivan Group, my experience at the time suggests that there was probably only one Y2K coverage person in the agency, there being so much debate over what could trigger defense and/or coverage.

A major national insurer had retained me as an advisor on coverage matters in mid-1997 to examine policy forms to determine where Y2K claims might successfully be triggered. While I will not go into the details of my analysis, I want to note that it was an important exercise in pointing out ambiguities in policy wording. This was my singular exercise of specific objective analysis of possible weaknesses in policy forms, which were never designed to address this arcane Y2K issue. My viewpoint remained as a rational insurance industry advocate for reason and balance on behalf of coverage intent and rating.

THERE WAS FOG EVERYWHERE, INCLUDING FOR INSURANCE MATTERS

This U.S. struggle between agents as well as between agents and insurers very late in the Y2K game is an important and crucial part of understanding how this global crisis was playing out at the time.

Most business journals were carrying articles, reports, and commentary on the likelihood of insurance coverage responding to Y2K problems. An August 1999 article by attorney Martha Bagley in *Women's Business* suggested, "consider conducting an internal audit comprised of reviewing insurance policies, contracts, and client needs." Advice concerning "strict compliance with policy requirements" and discussing "coverage questions with your representative"[35] still bewilder me these twenty-plus years later, since no standard insurance policy had any dictates regarding Y2K compliance requirements, nor had any insurance representatives the capacity to advise of courts' yet unsettled Y2K coverage interpretation response.

Once again, professionals in other fields of expertise offering insurance advice were frequently misleading consumers. Misleading and oversimplifying, both. In the same publication, editor Helen Graves, citing a recent newsletter of mine, misquoted me as saying that "it is unrealistic to expect that commercial insureds can collect more than pennies on the dollar considering the potential preponderance of costs to remediate Year 2000 technical problems."[36] In fact, I have no newsletter stating that; my May 1997 newsletter regarding Y2K issues reported that the nation's costs for compliance were predicted to be $600–$700 billion.

My email to the *Women's Business* editor must have noted that the entire U.S. casualty/property insurance sector totaled $280 billion. Note that reports of insurance reserves varied wildly, with almost $100 billion discrepancy in accounts. Reporting U.S. reserves at $260 billion was the *Insurance Litigation Reporter*.[37] Reporting "the cash reserves of the entire insurance industry in North America totaling only $380 billion" was NewsEdge.[38] Reuters New York reported, "Lloyd's of London estimated global claims $1T three times higher than the $356 billion in U.S. property/casualty reserves."[39]

Not quite enough to support the entire $1 trillion Y2K tab!* Collecting pennies on the dollar for insurance claims is customarily a consequence of insurer insolvency for legitimately covered claims. My position at the time was that Year 2000 claims were *not* going to be covered by insurance; uncovered for a variety of reasons, described in detail in this book.

More recognized names in the industry were publicizing proclaimant advice that, in my opinion, was deceptive. *National*

* See the *N. P. James Newsletter*, May 1997, reproduced in the appendix.

Underwriter carried an article interviewing William McDonough, senior vice president of risk management services at Marsh, Inc., in Boston. McDonough advised that insureds would be suing their insurers for remediation costs under "sue and labor" policy coverage clauses. I have excerpted it below for review.

During the runup to Year 2000, Marsh was posturing as an advocate for invoking coverage for its clients, a shrewd marketing move that depended on corporate America reasonably never having to test those coverage terms.

Sue and labor clauses require the following: 1) actions must be to prevent loss against a covered peril, 2) prevention of loss to covered property, 3) avoidance of an imminent loss, 4) reasonable expenses, and 5) expenses incurred must be for the primary benefit of the *insurer*. Each of these clauses had rebuttals for application of coverage.[40]

Sue and labor policy wording:

> In case of actual or imminent loss or damage by a peril insured or against, it shall, without prejudice to the insurance, be lawful and necessary for the insure . . . to sue, labor, and travel for, in and about the defense, the safeguard, and the recovery of the property or any part of the property insured.

Attorneys specializing in insurance matters were advising clients in much the same manner, using words like "Insurance companies are already re-writing their policies to exclude coverage for damages suffered as a result of the Y2K Problem" and "This does not mean that your existing coverage will not provide any remedy, however, and you may want to carefully examine your policy for sources of coverage and potential exclusions."[41]

Even with close friends and respected colleagues in these law firms, I consider advice such as this as helpful as suggesting that clients take a quick COBOL course to check their own computer code. Debate over the application of coverage among insurance experts was raging during the several years leading up to the new millennium. Recommending non-insurance professionals "review their policies" to determine coverage was neither helpful nor productive.

In a mid-1998 online article by the law firm Thelan, Marrin, Johnson & Briggs, LLP, a very harsh look at insurers, securities brokers, software vendors, and service providers described possible vulnerabilities in all *but* the legal profession. They suggested a complete review of all insurance policies, concluding: "If you don't have the right coverage, get it." For software providers: "[C]onduct a legal analysis of your contracts and licenses . . . place vendors on appropriate legal notice of your expectations." For securities investors: "[C]orporate management has a legal duty to shareholders to act responsibly to protect their companies."[42] This advisory document counseled the services of legal assistance throughout; both a substantial revenue producer for the firm as well as protection for the firm's own liability, having advised therein of legal counsel services.

A faxed copy of a letter from St. Paul attorneys, White and Williams, LLP, dated mid-November 1998, advised the insurer of the Year 2000 Information and Readiness Disclosure Act, passed into law on October 19, 1998. The letter itemized Y2K disclosure requirements with advisory notations, staff attorneys' names and contact information, as well as a full copy of the Act.[43]

THE FOG CLEARS: INSURERS ULTIMATELY RESPOND TO YEAR 2000 ISSUES BOTH INTERNALLY AND WITH THEIR AGENTS

Agents' associations, insurers, and insurance law specialists produced an excess of Year 2000 advisory material, most of which seemed then and now to be more self-protective than helpful. The following paragraph offers details on one agents' association Y2K compliance guide, which states on page 1 (italics in the original): *"Please do not hesitate to call your local independent agent with questions you have regarding you, your family, your business, your insurance coverage and Y2K compliance. Your independent agent can help!*"[44] The reason I saved four boxes of Y2K material was to later write about the national and international complexities of the issue.

"Countdown 2000," produced by the Independent Insurance Agents of America, Inc. (IIAA), in February 1999, offered an eight-minute video and two white papers, comprising a fifteen-page guide to Y2K compliance issues, including the suggestion that "your independent agent can help." A macro list of affected systems was followed by a what-could-go-wrong list, followed by a list of insurance policy areas (notable for the absence of Workers Compensation on the list), followed by a list of potential losses (Workers Compensation included), followed by a list of standard exclusions, and so on. Each list ended with a "*please note: this list is by no means all-inclusive*" and "*Ask your independent agent to explain these.*" Nothing like having one of the organizations to whom I paid annual dues let all my clients know I have the answers, even while national insurance companies were continuing the struggle with Y2K coverage questions. The short video displayed local agency personnel on computers with advice on both hardware and software Y2K compliance. The video ended with a lengthy disclaimer, effectively disclaiming all advice just given.[45]

With very short paragraph headings like "Is the Y2K problem fact or fiction?" "What is the Y2K problem?" and "What will be affected?" the document continues with "What is Y2K compliance?" stating: "There is no standard definition of what constitutes Y2K compliance. Individual companies have had to set their own definitions and develop their own strategies to become Y2K compliant." Further in the document is advice about protecting "fixed systems from *contamination*" (italics mine).[46] Another reminder that this was February 1999 and a document advising insureds to seek answers from their independent insurance agent!

Surely the IIAA knew then of the ongoing intense debates and conflicts over complex coverage response to Y2K issues.

Publishing a reasonably lengthy document seemed intended only to both disclaim any responsibility as well as to transfer impossible obligations to its member agents. It reminds me of the cartoon caption: "*If anyone has any questions on the presentation, ask Frisbee here; he made the copies!*" There began to be more and more such documents being offered entirely for self-protection purposes, transferring responsibility anywhere and everywhere else. Added to the burden of compliance came the additional onus of self-serving advisory memos from both partners and adversaries aiming liability elsewhere.

In these examples, it seems like the trillion-plus-dollars litigation costs might have been on target!

A more responsible insurer sent a letter to all agencies in mid-July 1998. Royal & SunAlliance, a London-based consortium of fourteen insurance companies, acknowledged that they had "invested significant resources" in potential Y2K exposures, outlining their twofold responsibilities to their customers: eliminate uncertainty and offer Y2K coverage options. "Some carriers consider Y2K exposures completely uninsurable and others are remaining silent as to their coverage intent. We consider both of these options to be irresponsible."[47] Nowhere did R&SA suggest liability for answers belonged to their agency partners.*

Leading up to the millennium turn, the other major agent's association was PIA. Its mid-1998 publication *Year 2000 Survival Manual*'s "Insurance Coverage and Client Issues" on Y2K offered much more thoughtful advice on Year 2000 coverage issues. PIA first covered the range of "carrier reasoning," describing opinions regarding coverage: the proximity of duty to maintain, consequential losses involving business interruption, CGL (commercial general liability) coverage being "all-peril" coverage subject to exclusions, and comparisons to pollution coverage. PIA argued against the "all-peril" position and against comparisons to

* Note: Royal's American subsidiaries stopped writing policies in 2003 due to its massive 9/11 losses, among others.

pollution, concluding that looming litigation might exceed all prior issues. The matter of Y2K policy exclusions was noted with both pro and con arguments; pro being the need for a clear insurer statement that Y2K coverage does not exist, and con asserting that the addition of late 1998–1999 policy exclusions suggest coverage existed on prior policies.[48]

The tort bar's position on the matter of late Y2K exclusion clauses on liability policies was as anticipated; coverage would be assumed to exist for Year 2000 problems on basic CGL policies—else, why would the need for exclusions exist? Xerox's counsel, Terry Budd, expressed exactly that when stating that insurers begging insurance commissioners to allow Y2K exclusions might end up haunted by their own poor planning. If existing general liability policy language did not cover Year 2000 issues, then why were exclusions needed, he asserted.[49] The insurance industry's two go-to manuals, FC&S and PF&M, carried no reference guide, industry-established practice, nor court decisions for assistance.

PIA's *Survival Manual* went on to suggest the wording agents should use in a letter to every carrier and wholesaler with whom the agency works. The letter requests a response on the carrier's specific instructions on coverage for Year 2000 losses as well as instructions on how the agency should specifically respond to insureds' Y2K coverage questions. In my opinion, that letter was very sound advice, leaving no agency responsible for determining carriers' response to Y2K coverage questions. I wish that I had paid more attention to PIA's advice and written to my carriers, although I knew then it would be unlikely that I would receive any definitive response. I had attended enough insurance seminars to know the industry was bracing for Year 2000 more than being proactive regarding Y2K coverage; again, there were no FC&S or PF&M manuals to offer guidance. Recall that my December 21, 1998, letter to my clients outlined the insurance industry and outside forces that would be determining coverage.

REGULATORS DICTATE COVERAGE

As fears over Y2K compliance increased, state insurance regulators were urged to step in, to protect citizens and businesses. I'll remind readers here that the McCarran–Ferguson Act of 1945 gave states the authority to regulate the "business of insurance." All forms of insurance were subject to individual state regulation as well as licensing on a state-by-state basis.

In February 1999, the Texas Department of Insurance (DOI) issued a bulletin prohibiting insurance companies from excluding or limiting Year 2000 liability coverage for a range of commercial businesses and institutions. The Texas DOI predicted that allowing Year 2000 exclusions "would lead to a chaotic situation in which businesses would seek retribution in a frenzy of civil lawsuits." As reported by the *Insurance News Network*, prohibiting exclusionary endorsements "will certainly keep insurance companies from running roughshod over businesses."[50]

Interestingly, in the same article, Connecticut regulators ruled that Y2K losses would not be covered unless specifically outlined in a policy. Massachusetts regulators ruled similarly, but supported federal legislation in the form of the Year 2000 Readiness Disclosure Act, which protected businesses (including insurers!) from litigation as a result of good-faith readiness disclosures.[51]

At the same time, debate continued over how any given Y2K-related loss would be regarded and/or covered. Insurers argued reasonably that Year 2000 was a business risk, not an insured risk. When challenged with the issue of automobile claim coverage should a Y2K issue show green traffic signals in opposing directions, insurers replied that the usual rules would apply. The collision's injuries and damages would be covered, with possible subrogation (collection of costs) from the liable party (most likely the responsible municipality). (A reminder here that government entities, from municipal to federal, were rushing legislation through to hold themselves harmless from liability!)

Arbella Mutual Insurance Company, a domestic insurer based in Quincy, Massachusetts, issued a letter dated August 16, 1999, "To Our Producers" stating that the Massachusetts Division of Insurance (DOI) would not accept or approve any Y2K exclusionary endorsements for personal lines policies (primarily home and automobile policies). The letter went on to state that Arbella's personal liability umbrella carried Y2K exclusionary language but did not explain any necessary changes as a consequence of the DOI directive. In addition, "It is important to remember that insurance policies are not intended to provide a warranty of products which are owned by insureds." The letter concluded with an assurance that claims staff could answer any specific questions.[52] In hindsight, this looks like a classic overview of what-if scenarios, some wilder than others, with an assurance of knowledgeable Y2K claims staff that no insurer should have been giving.

MEDIATION COUNCIL AND Y2K COURT

My article recommending the formation of a special council was published in the September 15, 1998, issue of the *Insurance Times*. (The article is reproduced at the end of this chapter.) I suggested the formation of a national body, the Millennium Mediation Council, including insurance, accounting, and law professionals to settle coverage questions and limits in advance; mediation before the fact for this unfortuitous Y2K event. I considered it necessary to determine claims limits within categories, uncovered areas of Year 2000 loss, attorney's fees, as well as specific, commonly defended areas such as fire or bodily injury. I concluded with:

> These areas can be quantified; we can advise our clients precisely; and we can be a valuable part of a global solution to Year 2000. Thus, we will be able to direct U.S. resources toward technical Year 2000 solutions and fail-safes. We can lead the EU and global insurance

communities in a rational strategy. And, we must urge our industry and government officials to support us in our efforts.[53]

SPECIAL COURT OR ARBITRATION PANEL FOR Y2K LITIGATION

The U.S. Chamber of Commerce recommended a special court be established to handle Year 2000 claims, designed to sunset after five years. In a December 1998 *Business Journal* article, Larry Kraus, president of the U.S. Chamber Institute for Legal Reform, was concerned with the complications and costs of finding blame for Y2K issues after January 1, 2000, and suggested a Y2K court be established to handle those matters. Seeking support from the National Association of Manufacturers (NAM) was disappointing, as NAM was supporting a legislated arbitration, limiting settlements to actual damages. The Information Technology Association of America's general counsel in the same article was doubtful that either a specialized court or arbitration panel would be workable, given the time delays in nominating and approving judges as well as the complications of Year 2000 cases.[54]

Robinson & Cole, LLP, also referenced the U.S. Chamber's proposal of a special court in their April 1999 "Update" newsletter. Noting the importance of alternative dispute resolution (ADR) and nonbinding negotiation and mediation, Robinson & Cole identified CIGNA, Bank of America, General Mills, McDonald's, and Phillip Morris as pledging those alternatives to Y2K disputes.[55]

Basically, these various proposals to direct efforts toward fixing Y2K computer issues instead of misusing resources on litigation pointed to the very real concerns for successfully meeting the technology challenge. In my commentary (in this chapter) of August 12, 1998, I concluded with:

This, singularly, like no other issue in history, is not only an opportunity but a necessity for professionals to work together to direct every available asset toward solving the Year 2000 conversion. Courts must cooperate, the bar must direct its efforts toward bringing its clients into compliance without litigation; congress and legislatures must redirect from municipal immunity to public assistance; while the public and private sectors join hands to solve the problem, short and long term.[56]

ABA 8/1/1997 NPJ address to ABA Annual Meeting: What is offered for Year 2000 Coverage now?

A.I.G., the world's largest insurance group, has developed what at this time is probably the definitive model for Year 2000 coverage. It is basically a **funding mechanism** for Year 2000 claims—thus, a $100,000,000 coverage limit is rated at **60–85% premium**. Arguably, the insurer maintains, **a $25,000,000 risk acceptance** is significant. Audits at a **$50,000 cost are mandatory, as AIG partners with its insureds** to prepare for Year 2000. The $50,000 is charged against **"Loss Expense**." Then, after all Year 2000 losses have been paid against that policy, the insured receives **premium back proportionate to its loss ratio**. Again, this is a funding mechanism and is one of only a handful of mid-1997 coverage options.

Johnson & Higgins, Marsh & McLennan, has developed a Year2000 policy with Zurich-American, Lloyd's, Reliance, and others, reportedly offered at more usual premium levels. As with AIG audits are mandatory, and costs not insignificant.

The entire **industry capacity**, however, is **only a fraction of the estimated $600–$700** billion cost to fix. **Insurance, surprisingly will most probably not be a viable place to turn for Year 2000 relief.**

YEAR 2000 INSURANCE COVERAGE PROGRAMS

Relatively few Y2K technology software remediators and consulting houses were willing to comment to reporters on their liability insurance coverage for Y2K compliance as the fourth quarter of 1999 closed. Most commented that their contracts with clients restricted damages to some small multiple of the contract price as well as a time element for corrections before penalties paid. Many stated that onerous liability requests from clients resulted in them walking away from the work.

In its mid-1998 publication, "Y2K and Your Insurance: The Issues," J&H Marsh & McLennan maintained that professional liability for technology-based enterprises was in a "nascent state" at that time.[57] I agree with Marsh's assessment, understanding that for decades there was less than a handful of technology risk insurers. That was primarily based on the highly publicized insurer losses in the 1960s when unintended policy coverage was invoked for the value of cancelled IBM leases when System/360 computers were announced. For more than two decades following, the only credible technology insurers were Chubb and St. Paul. It was not until the mid-1990s that other insurers ventured cautiously into the market.

American International Group (AIG) was the first company to offer Y2K coverage in early 1997, according to the May 1997 *Best's Review*. Called Millennium Insurance, offerings of both finite and blended programs were offered. Finite programs included premiums as high as 80 percent, with premium rebates depending on losses. Blended programs also had high premiums, 60 percent being the lowest over the term of the policy.[58] I am personally aware that Fidelity purchased a $100 million Y2K insurance policy for a premium of $75 million.

According to a later article in the June 8, 1998, issue of *Insurance Accountant*, AIG required an audit to "determine the overall track, credibility of and adherence to conversion plan." Premiums of 65 percent to 85 percent of limits were referenced.[59] That report tracks with my direct knowledge of the AIG policy, which I was told was offered on a retro plan, returning premium on a predetermined table basis depending upon losses. I had written in my "Commentary: The Y2K Problem and Insurance Coverage" in August 1998* that the AIG form had been withdrawn due to adverse selection (i.e., bad risk selection) from an announcement in June that AIG Executive Vice President Robert Omahne stated in mid-1998: "[Y]ou are not getting people who are planning for the future. You're getting desperate people."[60]

However, MSNBC reported in August 1998 that AIG was offering up to $50 million in limits for D&O policies Year 2000 coverage but declined to discuss cost. The report went on to say outside analysts reported coverage was "prohibitively expensive."[61] At the time, AIG was one of the leading insurers of corporate directors and officers, hardly in a position to painfully abandon its entire book for Year 2000.

Internecine as the financial world is often judged to be, the AIG product was developed at the request of and in conjunction with Minet Risk Services, a subsidiary of Aon. The AIG product differed from Aon's and Marsh's programs by offering two pricing formats, "funded cover" and "shared loss basis." The funded cover had premiums about 60 percent of the coverage limit, with up to 90 percent returned if there were no losses. The shared loss form had premiums 5 percent to 10 percent of the limit, with the policyholder sharing with AIG 50 percent for all losses, with no premium refund terms.[62]

Insurance companies had varied responses to Y2K coverage, most preferring to keep silent on their policy form responses. Royal & SunAlliance, in a rare departure from this norm, published a brochure in which they explained their coverage position. Directors and officers liability policies would grant coverage, which was appropriate and standard; some bodily injury and property damage protection; policy clarification endorsements; and availability

* Reproduced in the appendix.

for additional Y2K coverage.[63] (I do not have information beyond the brochure.)

Aon Risk Services of the Americas published a comprehensive 27-page document outlining the exposures of Y2K coverage line-by-line, from basic property and general liability to aviation, as well as professional errors and omissions liability. Aon, in conjunction with the reinsurer American Re, a unit of Germany's Munich Re Group, developed a millennium insurance plan called "Twenty-First Century" with limits to $100 million, qualification procedures and testing partners, and premium ranges. It also outlined a complex "finite risk" program written through AIG (too complex to detail in this book) with typical premiums of 80 percent or more of contract (coverage) limits with 5–15 percent insurer expenses. Insurer investment income of premiums is a consideration for the product, as well as a formula for premium returns to the insured depending on loss experience.[64]

Describing Aon's product in other publications as "Arm2000," limits were offered from $10 million to $1 billion with premiums ranging from 3 percent to 6 percent. Coverage options included business interruption, contingent business interruption, general liability, errors and omissions, and D&O liability.[65] Arm2000 also required auditing of coverage applicants, as well as validation of Year 2000 compliance.[66]

I include these to demonstrate the more complex programs designed for the very largest of clients' needs, those aimed at Fortune 1000 companies. Even so, Gartner Group's Year 2000 research director, Lou Marcoccio, was quoted in early 1998 as stating, "The insurance policies offered are totally nonviable."[67]

Year 2000 policies were "offered by J&H Marsh & McLennan, Minet & AIG, and Travelers Property Casualty Corp. among others," according to Thomas M. Reiter, Esq. of Kirkpatrick & Lockhart, LLP, in the *Journal of Insurance Coverage* of April 28, 1998.[68] In his 52-page document, Reiter suggested that the purchase of a Y2K specialty policy would be a tacit concession by the policyholder that coverage did not exist on the existing policy, a significant and hotly disputed issue in April 1998.* That was interesting, in contrast to Y2K policy exclusions being a concession on the insurers' part that coverage *was* granted by the existing policy.

Indeed, the horns of a dilemma. This is where one could conclude that in the end, it always comes down to dueling attorneys quarreling over the fine print.

The J&H Marsh & McLennan policy, named "2000 Secure," covered liabilities of wrongful acts, business interruption, contingent business interruption, and continuity services following a Y2K failure. A number of corporate due diligence processes were necessary to secure coverage.[69] Marsh's underwriters worked in conjunction with a newly formed Y2K auditing firm, 2000 Secure Audit Company, LLC, a mid-1997 joint venture of Ascent Logic and LeBoeuf Computing Technologies.[70] The 2000 Secure policy's lead underwriter was a Lloyd's of London syndicate, offering limits up to $200 million, and requiring an initial audit costing from an estimated $65,000 to several hundreds of thousands for a global concern.[71] Premium range reports varied significantly, from 2–6 percent of the limit to 10–15 percent of the limit. *BusinessWeek* quoted Marsh's managing director, William A. Malloy, as stating, "Premium paid would be up to $5 million for up to $200 million of coverage," which is a staggeringly low premium in comparison to the unknown Y2K risks, in my opinion.[72] So we might pause here to be reminded of the scope of discrepancies for Year 2000 products being reported to corporate America.

In the "Policyholders Guide" mentioned above, Thomas Reiter goes on to describe the AIG policy as a "risk funding mechanism" (finite risk) with large deductibles, requiring the insured to prepay between 65% and 85% of the policy limits, 90% of which is

---

* Reiter's colleague Matthew Jacobs, partner at Kirkpatrick & Lockhart, was quoted on July 7, 1999 ("Insurance Industry Decries the Y2k Claims—Update," *Newsbytes* via NewsEdge Corporation); "We feel strongly that there is coverage for these types of expenses." NewsEdge went on to report that the firm's numerous Fortune 100 clients were actively considering filing Y2K remediation cases against their insurers.

returned depending on the loss history." Reiter continues, "the J&H/Marsh & McLennan program, which has a non-returnable premium, is predicated on the insured's passing in quarterly Year 2000 audits conducted by consultants of J&H/Marsh & McLennan at a reported cost of $35,000 for each audit. The AIG product will also require expensive audits."[73]

In direct contrast to Reiter's "Policyholders Guide," a December 1997 article in *ComputerWorld* identified the Marsh 2000 Secure insurance policy as offered in conjunction with AIG and Minet, Inc., in London, not Lloyd's. Dominic Davison-Jenkins, vice president of risk consulting at Minet, said the policy would return between 90 percent and 95 percent of premiums, minus underwriting expenses and administrative costs. The lengthy application required for coverage would be reviewed by Arthur Anderson & Co. and AIG prior to binding coverage, according to Wendy Baker, senior vice president of Minet Reinsurance in New York. Premiums were estimated at 50 percent of the limit, with the policy period being from binding until January 1, 2000. Bob Cohen, vice president of the Information Technology Association of America (ITAA), representing a reported 11,000 members, warned that insurers would be very selective in issuing the Y2K policy.[74]

America's insurance capitol is arguably Hartford, Connecticut. The *Hartford Courant* ran a feature piece on September 22, 1997, on Y2K insurance availability, including the AIG and Marsh programs, as well as citing Travelers Insurance Company as planning to "offer some form of Year 2000 insurance in about six months." (The Travelers program never materialized, to my knowledge.) The article went on to quote Marsh's managing director Thomas Ruggieri as stating that no Marsh policies had been sold to date, while a dozen applicants were currently being audited for underwriting acceptance. Similarly, Marr T. Haack, USF&G vice president of Technology Marketing Group, stated that "the company is likely to offer a specific millennium product or enhancement to existing coverage." Again, I never saw a USF&G Millennium product offered in the market.[75] Connecticut-based General Electric's spokesperson Bruce Bunch was quoted as saying "We see it as a software fix, not an insurance fix." A Gartner Group analyst said, "First, it's practically nonexistence [*sic*], and it seems pretty ridiculous to buy insurance . . . instead of dealing with the problem."[76]

An interesting opinion in an early 1999 *Kiplinger Washington Letter* by the Kiplinger Washington editors to clients, offered that "Y2K insurance is available but not worth the very high price," going on to say that coverage comes with stringent requirements and inspections. "If you can qualify, you probably don't need it."[77] William Kelly, then senior vice president at J. P. Morgan and president of the International Federation of Risk and Insurance Management Associations (IFRIMA), citing insurance policies with a $20 million premium price for coverage of $200 million, commented, "They are heinously expensive."[78]

Kiplinger's conclusion was not exclusive to Year 2000. Similar observations were made by insurance executives, citing a *Time* magazine report that complex policy terms in cyber-liability (internet) insurance precluded actual coverage for any claim. That singular statement made in *Time* likely delayed the wide acceptance of cyber insurance for over a decade. Pollution insurance underwriting also requires extensive engineering tests to determine that a site's pollution exposure is completely known. Regarding Y2K insurance, while the cost and availability were consistent and widely reported, Kiplinger's editors and other experts' conclusions about need did not appear to be evidence based.

With Y2K insurance costs so widely reported and often criticized, my own approach was to urge my clients to spend their valuable resources on remediation, rather than any insurance product. It is advisable for all operational matters to address the root exposure and risk rather than rely on insurance proceeds to defend costly and damaging negligence litigation.

A non-insurance market, the American Bankers Association (ABA, not to be confused with the better-known American Bar

Association), issued press releases about its ABA-sponsored insurance program for Year 2000 fidelity exposures. Offered without additional premium, this limited coverage was for three months at a $500,000 limit, with higher limits and times being offered with premiums. Prompting the coverage need in part was the exposure of greater volumes of cash being required by banks leading up to the millennium. (Bank fidelity insurance differs from general commercial fidelity coverage in that it covers broad perils of mysterious disappearance, unexplained disappearance, and misplacement of cash, in addition to risks of robbery, burglary, and larceny among comprehensive coverage needed by banking institutions.) Of importance here is the need for industry-sponsored national insurance programs to offer Year 2000 coverage to its insured members. It is clear that associations put pressure on sponsored programs for needs and demands of members.[79]

CHAPTER 4 RECAP

- Coverage disputes involving "fortuity" (foreseeability), "late notice" of claims, "occurrence," "tangibility" of data, "peril," and other definitions for insurance recovery of Year 2000 computer programming remediation costs are detailed with various opinions.
- Fears of government declaration of the Y2K issue as a "virus" or "force majeure" to invoke insurance coverage funding for remediation costs is posed to point out Washington legislators' possible interference attempts.
- "Sue and labor" is discussed with the imaginative coverage issues and disputes for invoking coverage under a clause originally intended for marine risks.
- Corporate directors and officers liability is included with respect to the likelihood of liability as well as corporate D&O policies coverage renewal issues.

- The challenges and regulatory acceptance issues by state divisions of insurance of Year 2000 insurance policy exclusions are discussed. The dilemma of the tort bar's assertion of coverage being assumed under prior policies if Y2K exclusions are issued on renewal policies leading up to January 1, 2000.
- Insurers remained silent on coverage for Year 2000 remediation costs as January 1, 2000 approached, while agents and insureds awaited answers. Insurers, attorneys, accountants, consultants, and others were referring coverage questions to agents, as attaching liability to agents, who were equally awaiting answers.
- My Millennium Mediation Council recommendation is included, with other voices supporting special courts and arbitrations panels for Y2K dispute resolution as an alternative to expensive and public courtroom litigation.
- Special Y2K insurance policies written for large national and international clients, their terms, costs, and opinions of value are summarized, noting the puzzling scope of products and pricing.
- Conflicts between casualty/property total insurers' reserves of less than $400 billion in light of an expected $1 to $1.5 trillion expected tort bar litigation costs and damages are frequently referenced to point out a significance and magnitude of tension.

N. P. JAMES INSURANCE AGENCY
33 BEDFORD STREET
CONCORD, MA 01742
TELEPHONE (978) 369-2771 FAX (978)369-2778
NPJAMES@COMPUSERVE.COM WWW.NPJAMES.COM

COMMENTARY 8/12/1998

THE YEAR 2000 PROBLEM AND INSURANCE COVERAGE

The mere magnitude of Year 2000 problems creates a staggering obstacle when contemplating recovery of Year 2000 computer related failures through insurance claims. The total U.S. insurance industry reserves are reported to be $280 billion. With the worldwide costs of Year 2000 fixes estimated at $600 billion and litigation costs climbing to estimates of $1.5 trillion, a myriad of damage and liability claims simply cannot be expected to be turned over to insurers for defense and reimbursement.

It must be noted that AIG, the only insurer with active Year 2000 coverage policies in the United States, has closed the market from this date forward due to adverse selection.

This, singularly, like no other issue in history, is not only an opportunity but a necessity for professionals to work together to direct every available asset toward solving the Year 2000 conversion. Courts must cooperate, the bar must direct its efforts toward bringing its clients into compliance without litigation, Congress and legislatures must redirect from municipal immunity to public assistance, while the public and private sectors join hands to solve the problem, short and long term.

N. P. JAMES INSURANCE AGENCY
33 BEDFORD STREET
CONCORD, MA 01742
TELEPHONE (978) 369-2771 FAX (978) 369-2778
NPJAMES@COMPUSERVE.COM WWW.NPJAMES.COM

August 24, 1998

Guest Editorial
Insurance Times

via fax 617-292-0111
email instimes@tiac.com

A Modest Proposal
A Millennium Mediation Council

Having just returned from London, where the EU insurance community asked for a strong U.S. lead on Year 2000 coverage concerns, I am offering a modest proposal which breaks all our traditional business paradigms.

Just a year ago, on August 1, 1997, I addressed the American Bar Association Annual Meeting on the Year 2000 subject, concluding my early predictions on Year 2000 influences by asking the bar to join, as they have never before done, with their insurance and accounting professional colleagues in solving the Year 2000 problem for our clients.

Today it is a categorical imperative! Current estimates of this $600 billion global problem will exhaust twice over the U.S. estimated $280 billion domestic insurance reserves without even touching the $1–$1.5 trillion the bar expects in litigation costs and fees.

The EU community, without equivocation, concurs that U.S. courts will likely deny our Year 2000 exclusions. Exclusions denied assume coverage on the basic, unendorsed policy. Letters of clarification from carriers go a long way toward client understanding of coverage, but ultimately, no affordable independent commercial Year 2000 coverage exists in the states today. And, coverage areas have yet to be tested in the courts, which, in my judgment, will rival patent litigation costs of $500,000 pretrial and $500,000 at trial.

We still have an opportunity to solve this situation before Year 2000 arrives, and allocate every precious available resource toward the solution and away from adversarial dispute expenses.

A national body, *The Millennium Mediation Council*, including insurance, accounting, and law professionals must settle coverage questions and limits ahead of time; mediation before the fact for this unfortuitous event. Claims limits within categories need to be resolved, as does uncovered areas of Year 2000 loss; attorney's fees; and stated, common defended areas (fire, bodily injury). These areas can be quantified; we can advise our clients precisely, and we can be a valuable part of a global solution to Year 2000. Thus, we will be able to direct U.S. resources toward technical Year 2000 solutions and fail safes. We can lead the EU and global insurance communities in a rational strategy.

And, we must urge our industry and government officials to support us in our efforts.

Nancy P. James

5

CLAIMS AND LITIGATION

As with the next chapter on legislation, litigation arising from the Y2K computer technology issue has been exhaustively chronicled and analyzed by legal experts. What I am offering here is a late 1990s contemporaneous view and analysis of claims and litigation as I experienced it at the time from a risk and defense position. Among the thousands of documents I collected during the pre-Y2K 1990s, I have selected excerpts and data that typify the situations and considerations, as well as confusions and disputes of the time. It is my intention to offer commentary and opinion on some of the early and interesting cases. This chapter is written from the perspective of the litigation target, insurance; the anticipated funder of both offense, defense, and damages from Y2K problems. Ergo, it may be fairly interpreted as a less-than-objective view of the tort bar. Noted here, too, are more insights from Ramsay Raymond, psychologist and owner of advisory firm The Dreamwheel.

LITIGATION AND PERSPECTIVE

As mentioned in chapter 1, Ramsay Raymond stated, "This is a culture held hostage by fear of litigation" when referencing the pressures of Year 2000. And never more acutely was corporate America aware of this than when facing Y2K remediation challenges, with more and more focus shifting from actual compliance efforts to corporate protection from liability and litigation. At the time, some projected that for every dollar spent on remediation, two to three dollars would be spent on litigation.[1] With the American bar noting Lloyd's of London estimates of litigation expenses from Year 2000 to be a trillion dollars, anxiety over the cultural cost of this technological challenge was crippling.[2]

In Raymond's words:

> The deeper issue is what I see as a fundamental flaw or vulnerability in our legal system is the reliance on (disproportionately enormous) sums of money as the accepted form of punishment or of redemption for injuries sustained. This allows greed rather than justice all too easily to become a central player and motivator for litigation. Admittedly an efficient method of measuring recompense, it often does not serve the purposes of justice, which should not be limited to punishment but aimed at the protection and redemption of individuals and the social order. The huge sums given in civil suits are OFF THE WALL. You can bet it isn't the plaintiff at work here driving up a rationale for the extraction of such huge sums. You can bet none of this would be possible—suing professionals and huge corporations if there weren't the financial pools of insurance companies, corporate investments, etc. Perhaps it was appropriate that the woman won an award from McDonalds for the severe scalding of her leg, but the amount had no relation to the problem. The judges themselves no longer seem to have perspective. The costs of awards are passed on to the common person. Huge awards have to come from somewhere and it comes generally through insurance companies, which drive up the price for the common person—both health and liability.[3]

Raymond continues:

> Public Trust: A healthy society functions on the basis of trust. "Good faith" is the premise of any grouping of people; All exchanges are based on some degree of trust; Your interest is my interest; I will do my best to give you what you want, and I will receive what you give me in good faith. This takes place in billions of transactions large and small through each day and is the very fabric that makes social and financial exchange possible. In a civilized society, litigation exists to handle the exceptions and accidents that break the trust; litigation serves to restore order, to send warning signals that the breaking of trust will not be tolerated, or in the case of injury, that good faith restitution will be provided as feasible. Litigation is intended to restore and maintain the social fabric and serve the common good.
>
> What has happened is that litigation has now become an avenue for financial gain, and is cultivating an atmosphere of distrust, antagonism, self-interest and, most dangerously, double-speak. The fear of litigation is the new "Big Brother" that Orwell spoke of, the invisible **fear** that tyrannizes small and great. The art of double-speak is emanating not simply from the spin-doctors of media advertising and political competition, but in the courtroom. Lawyers are congratulated for winning, not for what justice they served—and often regardless of the facts, as in the case of O. J. Simpson. Professionals and corporations of every stripe are hyper-cautious. Doctors are afraid to take risks, people are reluctant to accept positions on Boards because they bear legal responsibility for corporate wrong-doing, politicians are reluctant to speak the truth to the public who cannot steer well if leaders aren't naming the realities we face.[4]

WHERE LITIGATION WAS HEADED

Battle lines between insurers and attorneys were drawn and expanding rapidly in 1997. Attorneys were frequently declaring that insurance would respond to Year 2000 litigation. Legal publications devoted to Y2K insurance coverage were looking toward a "litigation explosion."[5] Describing "Coverage Meltdown," attorney David Brenner outlined four elements: fortuity, liability over a period of years, good faith disagreements over policy terms, enormous monetary exposure. Brenner then added a fifth—"the creation of institutions which thrive on coverage battles."[6] Larry Eisenstein, a Washington, DC–based attorney, stated that "[b]usiness and product liability insurance is expected to cover the costs related to litigation."[7] Many experts estimated legal expenses would be about half of all Year 2000 claims costs.

Y2K expert Capers Jones, was quoted in the online "Year2000 Information Network" report: "[T]he end of the 20th century is likely to be a very hazardous time for many executives, and for almost all software executives." The network of Year2000 specialists, based in Ontario, Canada, published their report in December 1996, relatively early with such dire Y2K warnings: "Victims Will Create the Newest High-Tech Growth Industry—Litigation." Jones, whose work I cover in chapter 7, "Issues, Problems, and Litigation Solutions," was quoted at length regarding the list of possible litigation opportunities.[8]

Insurers and technology associations were expressing the opposite, however: "I think it is somewhat wishful thinking by some lawyers," stated Harris Miller, president of the prestigious Information Technology Association of America. "[S]uing each other is not going to fix the problem" he said.[9]

By early 1998, Lloyd's had reportedly received its first two large claims of over $1 million each. Nevertheless, Lloyd's did not predict claims levels as high as asbestos or air pollution, reported the *Insurance Accountant* in March 1998, quoting Martin Leach, spokesman for Lloyd's of London, and Matthew Jacobs, Esq. of the prominent Washington-based insurance litigation firm Kirkpatrick & Lockhart.[10] Contradicting that Lloyd's statement, Y2K reporter Scott Kirsner, in a lengthy article for *CIO Magazine* in May 1998, referenced Lloyd's estimating Y2K litigation costs worldwide to reach $1 trillion.[11] (Note the distinction between *worldwide* estimate to the oft-quoted *U.S.* estimate of $1 trillion litigation costs.) The $1 trillion litigation cost was reported and repeated in numerous publications in the United States and Western nations.

The same *Insurance Accountant* article was referenced as "raising eyebrows" when suggesting their corporate clients review insurance policies to determine coverage.[12] Again, contradicting that, Kirsner's article for *CIO Magazine* specifically advised technology chief information officers to involve their corporate attorneys in every phase of Year 2000 compliance and remediation, stating that most lawyers were skeptical that general liability policies would respond to Y2K issues.[13]

Kirsner interviewed a number of attorneys in representative parts of the country, noting advice regarding litigation; corporate attorneys are as averse to technology as IT professionals are to legal issues, the downside of suing mission-critical vendors and conducting legal audits regarding Y2K. Kirsner also reported on the growing number of pre-paid software update service contract vendors charging extra for Y2K updates, as well as the need for overdocumentation heading into the millennium.[14] Attorney Jeff Jinnett advised clients to review IT vendor contracts for existing obligations to cure "defects," "bugs," or "viruses," suggesting that Y2K might be characterized as one of those for the purpose of remediation work already agreed to.[15]

Insurance proceeds were widely seen as a plausible resource for Year 2000 remediation reimbursement. Insurance publications were fanning the flames of alarm as readily as other industries. Rich Huggins, in an article republished in a local Massachusetts agent's association publication, commented: "Of course, while the threat of litigation reduces candor and limits effective mitigation and contingency plans, it does help to focus them and motivate action as the officers and directors gradually understand to the organization's risk and their own liability for due diligence."[16] *National Underwriter* offered the opinion of the College of Insurance New York Associate Professor Alfred Jaffe: "As many as 20 percent of all companies will go out of business due to Year 2000–related problems."[17]

The Round Table Group, founded in 1993 using 750 experts from academia and think tanks, offered consultation services to clients, including law firms. It had established a Y2K Litigation Research Group for law firm clients designed for collaboration on customized research services and expert witness testimony: "RTG brings together these attorneys with the Y2K experts, quickly and efficiently, so that our clients can concentrate on winning cases rather than finding experts."[18]

"Y2K Legal Games Begin," by Lynda Radosevich in *Info-WorldElectric*, described the many Y2K issues facing IT managers, including a new facet of working with their corporate attorneys.[19] Mock trial scenarios were being performed for participant jury members, concluding in mixed verdicts and split juries, although most juries voted for conviction. An interesting note: these audience juries were later found to have voted more on performance than on facts, which happens in real courts.[20] Lessons learned.

Boston's *Mass High Tech* newspaper ran a Year 2000 advisory article by local attorney Joseph Blute of Mintz, Levin, Cohn, Ferris, Glovsky & Popeo. Blute stressed how important insurance would be to "potentially cover the problem, should the compliance efforts prove unsuccessful." Negotiating multiyear, noncancelable insurance policies, elimination of Y2K exclusions, placement of Year 2000 specific coverage, and a detailed understanding of all coverage and exclusions was advised.[21]

Even today, I can't help but read that and say, "good luck!"

Of course, both the insurance and legal sectors had little assurance of the ultimate application or scope of coverage on existing policies, nor, clearly, did they know the market for specialty Year 2000 policy coverage or cost at that time. But the pressure to transfer liability from the legal profession, which had drafted and executed those Y2K-vulnerable contracts and agreements, to insurers and their agents was well underway.

As insurers braced for Y2K litigation battles, they were relieved by the March 1999 U.S. Supreme Court case *Kumho Tire v. Carmichael*, holding that there is no distinction between "scientific" knowledge and "technical" or "other specialized" knowledge, applying an earlier 1993 *Daubert* case finding that expanded expert testimony. Looking to court rulings on matters of expert testimony on testing, peer review, known or potential error rate, and scientific community acceptance offered support to the insurance community. The complexities of software remediation for Year 2000 would challenge juries, so the rulings on both *Kumho* and *Daubert* affirmed a high standard of testimony to help eliminate the misuse of both junk science and mercenary expert witnesses.[22] We might pause here to note that limited Year 2000 compliance and remediation technology resources were being transferred to corporate legal issues, either out of fear, necessity, or both.

Software Magazine takes credit for the first (and they said, the best) creation of a timeline managing Year 2000 projects, "*Year 2000 Survival Guide,*" published regularly from mid-1996. In the October 1998 guide issue editor John Kerr noted in his letter to readers that buried in a Gartner Group's summer 1998 report was that 50 percent of all companies were not planning to test their Year 2000 compliance code. Kerr added a quote from legal expert Steven Hock: "If there's one thing juries understand, it's testing—including testing of all the products you buy."[23]

A May 1998 article cited Norton Utilities as one of the first U.S. software companies to be sued for alleged Y2K defects. Norton, based in Cupertino, California, had support from the California State Assembly's AB 1710, "a bill that would indemnify computer companies, and by association, their insurers, for defects relating to the Y2K virus."[24] By August 1998, six lawsuits had been filed against software companies including Intuit, Inc., and Symantec Corp., alleging that software that was not Y2K-compliant had been knowingly sold. Giga Information Group's estimates of $1 trillion in litigation expenses was then being compared to the magnitude of tobacco lawsuits.[25]

A similar *Boston Herald* story in mid-1999 titled "Lawyers Hope For 'Bonanza' Clinton Veto Could Bring" began with a quote from an attorney testifying to Congress indicating that Year 2000 litigation could be "the biggest litigation wave our country has ever seen." That sentiment was also repeated and confirmed by more sedate press in less dramatic terms. The article was referencing the proposed House and Senate bills limiting Y2K lawsuits that lawyers were hoping President Clinton would veto. Referencing *Lawyers Weekly USA* as source of the "bonanza" comment, even for small law firms, publisher Thomas Harrison detailed the breadth of litigation: business-to-business breach of contract, personal injury suits, tenant-landlord suits, attorneys assisting clients with Y2K limitations, and legal malpractice suits.[26] Including legal malpractice was a remarkable departure from most lawyer-based advice directed at suing everyone *but* attorneys. DecisionQuest was identified in the article as concluding that juries were very sympathetic to plaintiffs after a two-year study sponsored by aviation insurers of a hypothetical Y2K-related plane crash. The article went on to say the study's conclusion of jury awards which were adverse to insurers caused the "airline insurers to announce they would cover all Y2K-related 'events.'"[27]

That counterintuitive conclusion must have confused readers and set up insurer-versus-insurer clashes. The article quoted insurance giant Marsh's Boston vice president, John Klepper, as stating "We're doing all we can now to prepare, so that we'll be as protected as possible."[28] That hollow statement in my opinion was likely a no-response response to Marsh's primary competitor Aon's empty posturing that no Y2K exclusions would be permitted on their client's insurance policies.

With the legal profession, which had heavily supported President Clinton's campaign, hoping for a presidential veto, the competitive business and political forces of the tort bar, insurers, and technology service providers were aptly summarized in the *Herald* article. Plaintiffs' attorney Leo Boyle was quoted:

> My view is that big business is crying wolf and using this as a scare tactic to get a limitation on people's right to sue. . . . The legal profession

> didn't create the problem of negligently designed (computer equipment). It's like saying there's a lot of flu going on, let's blame the doctors.[29]

Dylan Mulvin, assistant professor at the London School of Economics and Political Science, authored "Distributing Liability: The Legal and Political Battles of Y2K." In his abstract, Mulvin stated:

> In 1999, the United States Congress passed the Y2K Act, a major—but temporary—effort at reshaping American tort law. The Act strictly limited the scope and applicability of lawsuits related to liability for the Year 2000 Problem. This paper excavates the process that led to the Act, including its unlikely signature by President Clinton. The history presented here is based on a reconsideration of the Y2K crisis as a major episode in the history of computing. The Act, and the Y2K crisis more broadly, expose the complex interconnections of software, code, and law at the end of the 20th century, and, taken seriously, argue for the appreciation of the role of liability in the history of technology. There were 18 Y2K legal cases in the United States as of January 2001.[30]

It is always interesting to get a glimpse of the EU's contemporaneous perspective on the events of the United States as the world watches and anticipates America's lead in global matters. Indeed, the role of liability in the history of technology is consistently affected by both the outcome of insurance litigation as well as legislation.

Predictably, risk managers were thrilled with the passage of the Y2K Act, signaling crucial protections against litigation. Insurance trade publications hailed the Act, reporting relief from corporate America. Lance Ewing, speaking for the Risk and Insurance Management Society of New York, was quoted in *National Underwriter* as expressing some ongoing fears of courts still being inundated with Year 2000 lawsuits, stating, "Sometimes I think litigation has replaced baseball as the national pastime."[31]

By mid-1999, reporting candid comments from teams from both sides demonstrated growing anxiety as the millennium turn stakes grew higher.

The American bar estimates grew between $1 trillion and $3 trillion in litigation claims for Year 2000 issues, consumer class action, D&O claims shareholder suits, computer systems malpractice and professional liability suits, product liability, fraud, breach of contract, and more.[32] Lloyd's of London was estimating $1 to $1.5 trillion.[33] Adding to the litigation fears was the *Kiplinger Washington Letter* to clients dated January 22, 1999, reporting on the U.S. Chamber of Commerce and other business groups supporting legislation to reduce punitive damages for Y2K issues. "Goal is to keep trial lawyers from going overboard on Y2K matters. It's a pipe dream," the editors noted, adding the chilling remark: "Trial lawyers are among the most generous contributors to Democrats."[34]

A lengthy article in *Insurance Litigation Reporter* in early 1998 predicted, "Whatever the actual effect of the Year 2000 Problem . . . two facts are certain: There will be litigation resulting from this computer glitch, and lots of it. There will be a first wave of litigation to shift Y2K losses to other parties, and a second wave of coverage litigation to shift liability to insurers."[35]

Both the insurance industry and trial bar launched numerous conferences to educate their professional sectors on risks, defenses, and claims scenarios. The trial bar was preparing both offensive litigation strategy as well as assertion of insurance coverage for the costs of defense and penalties. The U.S. insurance industry, with well under $400 billion in reserves in 1999, was understandably worried about another trillion dollars or so of Y2K defense and judgment expenses on top of any other category of losses. In a piece by Reuters New York via NewsEdge Corporation dated January 15, 1999, Lloyd's of London estimated global claims of $1 trillion, which was three times higher than then reported $356 billion in U.S. property/casualty reserves.[36]

It seemed that almost every law firm was eager to weigh in on the potential risks of litigation as well as the necessity to retain legal counsel to advise and audit enterprises for possible vulnerabilities internally and with business partners and vendors. In a special to the *Wall Street Journal* in November 1997, Christopher Simon reported:

> The New York law firm Millberg Weiss Bershad Hynes & Lerach, known for bringing shareholder class actions, has set up an in-house committee of computer experts and lawyers to explore various legal actions if a crisis does occur. Possible targets of litigation, says partner Melvyn Weiss, are corporate directors and officers. Mr. Weiss says management may be responsible for failing to disclose the costs of fixing the problem to shareholders. "Stockholder[s] could be blindsided," he says.[37]

Simon's article noted that even while Y2K threatened to cause chaos in the business world, lawyers saw it as an opportunity. Corporate attorneys were urging clients to write warranties into their software and IT contracts as a hedge against Y2K interruptions.[38]

Maria Recalde, Esq., in a special issue of *Women's Business*, recommended development of legal theories of Y2K liabilities for the technology products sector, including determination of legal obligations for Y2K-compliant upgrades. Costs for both compliance and for failed compliance should be assessed, she noted, adding that contracts going forward should contain warranty disclaimers or limitations, limitation of damages, and Year 2000 compliance language. Recalde offered valuable advice, including a robust corporate record retention policy with specifics for Y2K efforts; a coordinated response to Year 2000 readiness inquiries should be designated as "Year 2000 Readiness Disclosures under the Year 2000 Information and Readiness Disclosure Act (Pub. L. 105-271, 112 Stat.2386) in order to obtain the protection afforded by the act." Five steps for comprehensive Year 2000 internal and external efforts included record retention, contingency planning, legislative and regulatory monitoring, a Y2K expense tax strategy, contract and, lastly, to "address insurance coverage issues."[39] Recalde's outline did not suggest her firm, Burns & Levinson, LLP, could provide support for any or all of her Y2K compliance steps.

In the contemporaneous documents I have, I read of no attorneys offering insurance contract interpretation to clients.

In the same issue of *Women's Business,* public image consultant Deborah Chiaravalotti recommended a series of communications strategies to assure stakeholders of corporate Year 2000 readiness. Strategies such as a "standby statement," media kit, public outreach program, development of a Y2K regulators guide, and keeping perspective on the hysteria around Y2K were outlined, with corporate communications being ultimately the primary objective of public confidence, according to Chiaravalotti. Suggesting the possibility of a national disaster-like media blitz on January 1, which she cleverly dubbed "Katastrophe," coverage would mirror those of natural disasters, including runs on banks, food hoarding, and gas station lines. Like many advisers for Y2K, she suggested a daunting list of outreach programs to educate stakeholders; "special newsletters, direct mail to customers and presentations to public and private audiences" as well as "develop a guide to Y2K for regulators or other bodies influencing your business . . . dedicate a telephone line to Y2K inquiries," and "post and regularly update information about your company's Y2K preparedness" [on your website].[40] While maintaining a public image in advance of the millennium turn was important, advising such a comprehensive program specific to Y2K readiness seems over the top now, if not at the time. Distractions from corporate core business goals were all too easy to recommend at the runup to the new millennium for many reasons.

Attorneys offered varying opinions of where and how litigation might arise. Rhode Island attorney Mark Freel suggested that "Most commentators appear to agree that contract-based remedies are likely to provide the greatest source of relief." Warranties of merchantability and fitness for a particular purpose were logical approaches for litigants.[41] I certainly agreed with Freel's approach, recommending in the late 1990s to attorney friends and associates that same specific Y2K response language be incorporated into their vendor contract wording. Interestingly, Freel went on to note that the computer field at the time had no clear industry standards of performance; thus, "malpractice" claims had already been rejected by most courts. Again, harking back to contracts, Freel recommended that contracts contain language specific to professional standards and warranties for *services* rather than *goods*.[42]

I found during my research that such comments on standards of malpractice in the computer field had been noted from various quarters. I was surprised that most insurance professionals were not actively offering risk management, risk transfer options, and policy-to-contract compliance advice to technology-based clients. Professional malpractice insurance for the technology sector, both for manufacturing and service sectors, was available but little utilized by most main street brokers and agents. Most insurance personnel had business backgrounds rather than technology industry experience, and thus were untrained in assessing technology risks. Over time the insurance policy forms themselves made up for this lack of direct experience and expertise, providing detailed definitions and coverage explanations as well as brochures with claims scenarios for both brokers and clients.

Michael R. Cashman, a partner with the law firm Zelle & Larson, LLP, in Minneapolis, in a May 1998 national satellite broadcast to the CPCU Society reported in *The Standard*, stated that there were predictions "that 20 percent of the businesses in the country will go out of business as a result of the Year 2000 problem." In the same broadcast, Jeff Jinnett, counsel for the New York City law firm of LeBoeuf, Lamb, Greene & MacRae, added that insurers should be concerned about Workers Compensation claims, which have a tendency to spike when companies fail. Jinnett summarized by saying he "believes the country will be able to 'muddle through' the problem."[43]

By January 1999, Year 2000 problems had already begun to appear, and as enumerated above, insurers were increasingly nervous as the trial bar's excitement increased. Insurance professionals at a conference sponsored by the Washington-based International Insurance Council heard speakers refer to the possibility of a technological apocalypse. Insurance Services Office (ISO) chair Fred Marcon, addressing the gathering, said that with "the decline of economic protectionism, the privatization of national insurers in Latin America and the increasing competition in mature markets such as Japan, insurance companies face new uncertainties, risks and the need for new tools to compete successfully in a new environment." Other speakers referred to the $1 trillion in litigation costs as a "world and insurance industry paralyzed by litigation," offered by Peter R. Martin, business strategist with Jams/Endispute. Martin said that litigation should be the means of last resort, with business-to-business solutions, mediation, and arbitration being preferred and less costly solutions to disputes. This was at the same time that Congress was introducing legislation to ease liability and litigation with arbitration requirements and judgment limits.[44] We are reminded of those same sentiments by psychologist Ramsay Raymond: "This is a culture held hostage by fear of litigation" (noted in chapter 1), which is rarely stated in such stark terms.

As late as November 1999, David Reich-Hale in the *National Underwriter* reported that insurance experts warned that insurers are customarily regarded as the "bad guy" and litigation success would be difficult for policies with clear Y2K exclusions. In denying Y2K-related claims, "proximate cause" (meaning *originating* cause) claims issues would be raised. For example, if fire or frozen pipes were to occur as a result of Y2K utility failures, would the claims be covered? (Both fire and resultant water damage from burst pipes are routinely covered by most property policy forms.) This is the same Y2K "proximate cause" argument as Workers Compensation injury claims by employees required to manually take over assembly-line functions previously done by mechanical tools, which failed for Y2K issues.[45] Reich-Hale's observations of the numerous outstanding unknowns of the impending millennium expressed valid concerns from every sector.

MILLENNIUM RISKS

Very little was written on remediation for the millennium turn from a risk analysis perspective. *Software Magazine*, in its final Year 2000 readiness issue, ran a technology-based feature article on testing, including a section on risk-based testing. The opening sentence, "Testing is a form of insurance," outlined *cost of failure* and *risk-based testing* in strategies targeted to the software industry. Identification of high and low impact systems described a triage method within the larger remediation model.[46]

Confusion over whether a Y2K "malfunction" or "glitch" might actually have occurred was a hot topic of discussion among both legal and insurance professionals. Computers reading "00" as 1900 would be operating properly as designed and implemented, operating exactly as intended. Contrarily, any computer meant for functional application around the calculation of dates which could not handle "2000" may be construed as defective. Litigation of this very issue was reasonably anticipated.

In an article published in the *CPCU Journal* of January 1998, Joan Hartnett-Barry cited Emily Canelo, senior vice president of Zurich Reinsurance (North America), reminding readers that:

> Case law has shown that plaintiff attorneys in construction defect cases have successfully argued that the defect that caused the injury occurred when the building was built, even though the defect didn't manifest itself until years later when the injury happened. The same type of theory has been used in asbestos litigation and environmental impairment cases.[47]

I note this article from one of the insurance industry's most prestigious and widely read journals to point out the scope of claims theories leading up to Year 2000. Well before that time, in January 1998, I was of the opinion that two-digit year date codes were not an "inherent defect" in software programming (which I have described in the introduction); two-digit year indicators were universally recognized as acceptable and cost-effective. However, the widely accepted "inherent vice" argument, while encouraging to plaintiffs' bar, also urged the technology and insurance sectors to fix the problem!

The Massachusetts Association of Insurance Agents (MAIA) published an initial *Tech Talk* paper in October 1998 with detailed Year 2000 form and coverage analysis authored by policy coverage expert Irene Morrill.[48] Morrill did not offer her opinion of the six-digit versus eight-digit date field issue in her carefully detailed work, but guest author Rich Huggins did so in the following November 1998 *Tech Talk* Y2K article. Huggins's opening paragraphs for his "The Year 2000 Hazard" commentary stated: "The Y2K problem results from a shortcut used in many computers and microchips. Years ago, to conserve memory space, programmers developed the custom of using two numbers to record the year."[49]

This was a very close-to-home misinterpretation of the financial origins of date coding, like so many other arcane coding schemas. Huggins's characterization in a technical piece from the insurance side only added to the opinion supporting negligence in both systems design and programming. I point this out to demonstrate that misinformation came directly from a source who should have known better.

Attorneys representing insurer interests were also preparing defenses against coverage claims from plaintiffs' attorneys. Policy language on the ISO Commercial General Liability (CGL) policy was carefully detailed with explanations and defenses. Arguably, I regarded at the time—and still do now—that most of the liability claims would ultimately reside on the desks of technology software vendors and services.

In the late 1990s, there were still very few technology insurers offering Year 2000 coverage, and even fewer as January 1 loomed/approached. Policy forms were specific to each insurer, with few relying on ISO standard forms. Those forms, as I have mentioned in the previous chapter, identified "data" as *intangible* in the definitions policy section. Thus, the huge issue of tangibility was moot in most of the policies issued by technology specialty insurers to those software firms responsible for Y2K compliance work.

The reason there were so few technology insurers extends back to the mid-1960s when IBM presented a huge claim to its insurers when it introduced IBM's new 360s. Most IBM customers cancelled their existing leases, turning in their leased equipment for the newer 360 version. The lost lease revenue turned out to be an insured loss—better stated—not excluded from IBM's insurance policy. IBM's insurer paid millions in claims, resulting in reluctance on the part of other insurers to enter the technology risk arena for a number of years. As a consequence, there existed a void in technical understanding of Y2K computer issues within the insurance underwriting ranks.

Cautions regarding statutes of limitations governed by the Uniform Commercial Code (UCC) were offered by the Boston law firm of Sugarman, Rogers, Barshak & Cohen, PC, specifically the four-year statute of limitations on vendor products running from the date of purchase. The UCC statute is unaffected by buyer's knowledge of defect, and could easily expire before any Y2K defect or failure is known. Specific exceptions and necessary triggers were described. Legal complexities of Y2K claims for "negligence," "breach of contract," or other causes allowing for longer claim periods pointed out the intricate web of statutory legal maneuvering.[50]

CONTRACTING FOR Y2K TECHNICAL ASSISTANCE AND INSURANCE PROTECTION

Contracting for Year 2000 compliance assistance had some interesting alternatives to standard services contracts. My research for this book found a mid-1998 white paper by Gregory P. Cirillo, Esq., partner at Williams, Mullen, Christian & Dobbins, walking back his earlier recommended contract "risk allocation philosophy." Traditional contracts of warranties, financial incentives, and onerous economic remedies if obligations were not met, placed the economic risk of failure on the contractor. By 1998, that simply would not be acceptable to Y2K compliance contractors due to the realistic pressures for Year 2000 remediation services, then less than two years from January 1, 2000. "No leverage, no alternatives, no time" was examined in light of Y2K technology services being in very short supply, with vendors holding most of the contract negotiating power.

Except for contract work for the largest enterprises, businesses were offered "process-oriented" versus "product" contract advice, which should lead to greater Y2K compliance success if carefully audited and managed. "A process-oriented contract requires the contractor to commit to performance standards and personnel requirements, rather than focusing on the end product for fear of economic consequences."[51]

Cirillo offered a detailed alternative contract philosophy process versus product: Y2K contractors' personnel, experience, expertise, and best efforts. Arguably, Y2K contractors would be willing to accept such terms given that before commencing services, the contractor would have no way to assess systems, code, documentation, or time needed to review compliance scope and work. Y2K contractors have more control over the process than the outcome; experience and best efforts, not a guaranteed result, are primary concerns. Cirillo added that needing to contract for guaranteed success would come with a very heavy price tag. In addition, client ownership of all work product and licenses should be a part of the contract.[52] I want to include this description to point out this distinctive contract philosophy, which I saw and considered only during the Year 2000 run-up and rarely during these subsequent twenty-plus years.

It is important to note here that most hardware and software licenses have complex agreement terms, with ownership not customarily held by the user; software then and now is licensed, not sold. Software license contracts contain terms for conditions of use, trade secret content, third-party maintenance restrictions, as well as prohibited use terms, among numerous other terms.[53] As mentioned above, those problems involving license agreements and Y2K-compliance remediation became painfully evident when heading into the new millennium.

Cirillo's conclusion was the very same one that I had come to for Y2K services contracts. This is interesting as an important departure from standard contract term recommendations to a very realistic approach for securing Y2K software compliance services crucial to a favorable outcome, and by the same attorney who had authored more standard terms in early 1997. Cirillo's opinion was based on the certainty that those in need of Y2K-compliance help would have almost no contracting leverage with scarce Y2K remediation services. Again, I saw and considered this contracting philosophy only during the Year 2000 runup and rarely since.

With respect to the limited insurance products on the market, I remained unsure if recommending any available one for Y2K risks was prudent, either for clients, or as a requirement for Year 2000 remediation contractors. I became sure of that decision when I received a call from an attorney associate who was meeting with a New York client in early 1999. She asked me if her client's contract should specify Year 2000 insurance from the Y2K remediation party. My immediate reaction was to look at any outcome that might necessitate an insurance claim, and I responded that an insurance requirement was not the way to proceed. I recommended that the contractor's Y2K programming staff and management, by name, if possible, be detailed in the contract with specific dates and times, including December 31, 1999, for on-site presence.

My rationale: An insurance policy requirement only provided funding for defense and judgment in the event of system failure due to Y2K issues, and also inevitably suggested lengthy litigation, not the required Y2K systems compliance. Specifying personnel by name or title to be on site during the millennium turn would be the best guarantee of success. My associate agreed completely. Year 2000 insurance programs are discussed at length in chapter 4.

LAWSUITS

The first Y2K case filed in 1997 was against Tec-America by Produce Palace in Michigan, when Produce Palace's cash registers crashed, incapable of reading credit cards with expiration dates of 2000. That $5 million suit was settled out of court for $250,000 through mediation.[54] (An additional $10,000 was paid by the system installer, American Cash Register.) I have an October 1998 email from an attorney in the Washington, DC, area stating that she had spoken with the defendant's attorney, stating that the case was settled for "the mediator's recommendation of $260,00. They will not disclose whether any portion of the settlement was covered by insurance."[55] The *New York Times* reported Washington attorney David Nadler as disputing the importance of the Produce Palace suit because software date problems preceded Year 2000, saying "It's a lemon-law case dressed up in Year 2000 clothing."[56] This points out the disparate opinions within professions as industry segments grappled with the coming millennium.

Reported as the "First Class Action Settlement" by *Mealey's Publications* in October 1998, Software Business Technologies, Inc. (SBT), agreed to attorney's fees, Year 2000 software upgrades on current products, and discounts on Y2K-compliant software for older versions. The suit was brought December 2, 1997, by the law firm of Milberg, Weiss, Bershad, Hynes & Lerach on behalf of Atlaz International, Inc.[57]

The second settlement involved a class-action suit against Medical Manager for billing software that could not recognize or process dates after December 31, 1999. The settlement included free Y2K-compliant software upgrades.[58] Again, insurance claims were not a part of the suit or settlement.

RealWorld Corp's class-action suit on June 26, 1998, predated federal statutes, specifically the Y2K Act and the Year 2000 Information & Readiness Disclosure Act. RealWorld of Manchester, New Hampshire, provided Y2K compliance on only the current version of its accounting software. RealWorld's mistake was concluding that updating older versions of its software would be cost prohibitive; thus, patches were not offered to existing clients. Five hundred businesses ultimately joined a class action suit, resulting in RealWorld providing a software upgrade and support terms, and importantly, avoiding customer Y2K software accounting issues as well as the cost of litigation.[59]

Another suit, in what was reported to be based on the first legal liability decision in a Year 2000 case, involved a $3.9 million suit against ASE Limited over a five-year re-engineering contract executed in 1995 that had no mention of Year 2000 remediation or compliance. The arbitrator denied the claim because Y2K was not mentioned in the contract and ASE could not be held liable "as an afterthought."[60]

One of the most important cases involved Cincinnati Insurance Company. It was the first request for a "duty to defend" declaratory judgment involving an insurance company under a commercial general liability (CGL) policy. Filed on December 4, 1998, by Cincinnati Insurance Company against its insured, Source Data Systems (in 1998 owned by Keane, Inc.) and Pineville Community Hospital in Pineville Kentucky, Cincinnati requested the court declare that the insurer had no duty to defend. This was a very closely watched case because CGL insurance was, and remains, a basic foundation of every business insurance program. The precedent of finding Y2K coverage successfully under a CGL policy would arguably bankrupt the U.S. casualty/property insurance market; every insured enterprise in America would present its Year 2000 remediation expenses to its insurer for reimbursement.

Cincinnati Insurance asked the Federal District Court in Cedar Rapids, Iowa, to rule that their insurance policy did not require them to defend for Year 2000 noncompliance, as reported in the *New York Times*.[61] Pineville had sued Source Data Systems (SDS) for the replacement costs of their noncompliant Year 2000 system installed by Source Data in 1995 and 1996. Policies in question did not include professional liability coverage; the policies covered only the standard public liability perils of bodily injury, property damage, and personal injury (insurance contracts define "personal injury" as defamation, slander, libel, private occupancy, loss of business opportunity, etc.). Cincinnati's defense cited "an exclusion for any property damage arising from '[A] defect, deficiency, inadequacy, or dangerous condition.'"[62] The facts of the case clearly fell within "professional errors & omissions" coverage, which was not provided. This *Cincinnati Insurance Company v. SDS* case was being closely followed by the insurance side, with Martin Sheffield, an analyst with A. M. Best, being quoted in the *Insurance Times* as saying "This would be a pretty big precedent if Cincinnati Insurance's defense doesn't pass muster."[63]

The *New York Times* article reported that SDS denied that its contract with Pineville guaranteed Year 2000 capabilities, but if found liable, its insurer, Cincinnati, should cover. In a departure from most of the press reporting of this case, the *Times* article stated that Cincinnati Insurance agreed that "its general liability policies will cover some Year 2000 problems. But, the insurer argues, the cost of curing foreseen defects before there is any damage to individuals or property is clearly outside the policies' scope."[64]

Boston attorney Arlene Polonsky addressed the Cincinnati case in her late 1999 article for *Defense Research Institute*, clearly outlining the insurer's defense:

> [T]he Pineville action alleges causes of action based on fraudulent conduct and breach of contract. . . . [I]t does not allege "occurrence" as defined by the policies; it does not allege "property damage" as defined by the policies; it does not allege that any "occurrence" or "property damage" happened during the policy period as required by the policies; it does not seek recovery for sums paid as "damages" as required by the policies; it alleges a loss which was a known loss from the viewpoint of SDS; it alleges a loss that was in progress. . . . Exclusions for "your work," defects, deficiencies, performance delays and professional services were also tendered by Cincinnati.[65]

The *New York Times* clarified that one of the issues is an April 1999 article stating the "actual insurance policy at issue may not cover the crucial years in the underlying suit against Cincinnati Insurance's client," and went on to say that that alone may let the insurer off the hook.[66]

In its 242-page book, *Insurance Issues for the Year 2000* published by the Defense Research Institute, an organization for defense attorneys and in-house counsel, all of the objections to defense and coverage for Y2K were repeated. The issues of "occurrence," "property damage," sums paid "as damages," and no fortuity as a known loss, were specified for the Cincinnati suit and described in detail throughout the book, while noting the importance of the outcome of this first case. The court ultimately decided to await the outcome of Pineville's case against Keane (then owner of Source Data Systems) before ruling on Cincinnati's duty to indemnify. Cincinnati's duty to defend and related discovery in the meantime was ruled to continue.[67] Cincinnati had initially agreed to tender defense with a full "reservation of rights" letter (such a letter allows claim investigation, with a possibility of later denial of the claim, refunding insurer's costs expended if no duty to indemnify is found). This case was so closely watched because CGL insurance was, and remains, a basic foundation of every business insurance program. The precedent of finding Y2K coverage successfully under a CGL policy would arguably bankrupt the U.S. casualty/property market.

I have included unusual detail about this case due to its potential for ruinous consequences to the casualty insurance sector as well as because it was so widely publicized. As this was such a closely watched case from many sides, Cincinnati Insurance Company may have served as a damper on eager attorneys' litigation hopes, as layers of complex policy terms were invoked as defense.

With so much initial media attention to the Cincinnati case, it was surprising that no outcome was found in the thousands of documents I have leading up to and following the millennium turn. Having had the good luck to meet two Cincinnati Insurance Company executives at a cyber conference in early 2022, I reached out to them requesting details on the outcome of that case. The company's response to me was "no comment," even after all those years. Further attempts to reach out to the company have resulted in the same reply; "no comment." Some considerable online follow-up research into the originating case, *Pineville Community Hospital Associates, Inc. v. Keane, Inc.*, found the matter settled privately in April 2003 with the suit dismissed. Details were not readily available on any involvement or further litigation against Cincinnati Insurance Company by Pineville. After more than two years of searching unsuccessfully for the outcome of *Pineville v. Cincinnati*, I conclude that it, too, may have been settled within the *Pineville v. Keane* mediation;[68] Cincinnati perhaps accepting defense costs while being relieved of damages (there was arguably no duty to indemnify). Such private mediation avoids a catastrophic court precedent from a jury sympathetic to a community hospital and not sympathetic to Keane's insurer! As all of this occurred well after January 1, 2000, those interested parties concerned with insurance coverage for Y2K costs still did not know!

Another early case in Massachusetts against Andersen Consulting by shoe manufacturer J. Baker, Inc., involved a non-Y2K-compliant computer system installed in early 1998 by Andersen. Again, insurers nationally were intently watching that case, too, to see how insurance policy terms might be interpreted, worried that sympathetic juries could ultimately be determining terms. Andersen's defense asserted that its contract with Baker clearly defined a two-digit date field. Baker's complaint asserted that the $3 million cost for the retail software installed by Andersen should have been Year 2000 compliant.[69] Andersen cross-sued, requesting declaratory judgment to determine breach of contract and duty owed to J. Baker.[70] The suit was settled through mediation without payment to J. Baker; insurance was not involved. A joint press release published by Andersen Consulting News stated, "Following the mediator's review of the dispute, J. Baker re-evaluated its claims and is now satisfied that Andersen Consulting had met all of its contractual obligations to J. Baker."[71]

By December 1999 as the new millennium loomed, attorney David Tennant's article in *Mass High Tech* looked at Y2K litigation to date, focusing especially on insurance coverage claims. Lawsuits fell into a number of categories: Half of all suits involved charging for compliant software upgrades; one-quarter of Y2K suits were shareholder actions; six suits involved charging for hardware upgrades; five large suits were under "sue and labor" insurance clauses for expense reimbursement; three suits were for failed remediation/consultation; two were for systems failures due to Y2K; and the closely watched Cincinnati case described earlier.[72] Those statistics offer an important point to the conclusion of plausible insurance coverage actions.

THE IMAGINATIVE "SUE AND LABOR" AND "FORTUITY" COVERAGE ASSERTIONS

GTE (seeking $183 million in remediation expenses), Xerox, CVS, Unisys, J. P. Morgan, Stanford University, Nike, Kmart, The Gap, Microsoft, and the Port of Seattle, among others, sued insurers for their remediation costs under sue and labor insurance policy clauses. One of the determinations of sue and labor coverage was that the loss needed to be "imminent." Xerox in its 1998 annual report affirmed that Y2K remediation efforts had been ongoing since 1996. GTE in its 1998 annual report also noted Y2K efforts since 1995. J. P. Morgan posted its Y2K timeline dating from 1997. Such long-term actions would not fit the definition of "imminent."[73] By the end of 1999, Y2K claims totaled $700 million, with many more expected to come as insurers braced for a fight.[74]

Reported by *Newsbytes* on July 5, 1999, GTE Corp. sued five of its insurers for $287 million of the anticipated $400 million spending on Year 2000 remediation based on insurance policy wording covering expenses avoiding an "actual or imminent loss." GTE's preemptive suits against their insurers, including Allendale Mutual Insurance Co. and Allianz and Chubb Corp., were brought in advance of their insurers' responses to GTE's Y2K claims. An insurance industry source was reported to have called the suits "ludicrous," noting that such clauses, also known as "sue and labor" clauses, are intended for sudden situations in which a partial loss is deliberately undertaken to prevent a larger loss. (Dating from the seventeenth century, such clauses are common in marine insurance, where jettisoning a portion of cargo to save the ship from sinking would be covered by the insurer.) The imminent nature of the cause of action is paramount to invoking coverage, Year 2000 being a long considered event to the contrary.[75]

Sue and labor clauses are actually intended to protect the insurer, who promises reimbursement for expenses in efforts to protect property. American Insurance Association spokesperson Dan Zielinski said that the sue and labor clause was being "misapplied" for Y2K suits. Indicating firm legal ground for defense, Zielinski noted the lack of defect and unforeseeable events denying application of sue and labor clauses.[76]

An article in the *New York Times* of July 2, 1999, named all five insurers to include Affiliated FM and the two Chubb insurers, Federal Insurance Company and Industrial Risk Insurers. Washington attorney Robert F. Ruyak, representing GTE, indicated that insurance policy language spoke directly to the Y2K issue:

"[D]estruction, distortion or corruption of any computer data, coding, program or software" as well as the "sue and labor" policy wording. The article also cited the actuarial firm of Milliman & Robertson as estimating between $15 billion and $35 billion in U.S. claims and legal expenses.[77] This is a far cry from the bar's anticipated $1 trillion in litigation expenses, much of it from insurance proceeds.

By the end of July, insurance attorneys were taking sides on the merits of GTE's claims, as reported by *National Underwriter*. Specifics of the insurance companies in the suit in this closely watched case included: "FM Global Insurance Company (formerly Allendale Mutual Insurance Company), Johnston, R.I.; Affiliated FM Insurance Co., an FM affiliate also based in Johnston; Burbank, Calif.-based Allianz Insurance Co., Federal Insurance Co., Warren, N.J.; and Industrial Risk Insurers, based in Hartford, Conn." Coverage assertion was based on policy wording: "any destruction, distortion or corruption of any computer data, coding, program or software." All primary and/or excess policies also included sue and labor clauses. Attorneys for the Insurers' Year 2000 Roundtable, made up of 33 insurers and reinsurers, disputed coverage declarations, citing fortuity as well as asserting "Year 2000 remediation costs are ordinary business expenses—the ongoing costs of modernization and upgrading computer systems and manufacturing facilities." Plaintiffs' attorney, Andrew M. Reidy, considered *fortuity* to be where court fights would be, stating, "You can't insure what is a known *loss*, but you can insure a known *risk*." Reidy added that GTE, even while knowing Y2K was an issue, did not know its $400 million cost of remediation.[78]

Attorney Reidy suggested that individual state laws would be applied in the courts, and argued that claimants only need prove that they had not intentionally caused the loss for coverage to apply. In the client bulletin of Washington-based McKenna & Cuneo (Reidy's firm) Year 2000 cost recoveries were strongly stated to be covered in the absence of a Y2K policy exclusion, asserting that the burden of proof for denial of coverage was on the insurer. In addition, Reidy suggested immediate filing of claims to avoid later "timely notice" arguments from insurers. The Roundtable disagreed, of course.[79]

Intense assertions of insurance coverage response for Y2K claims were heating up as January 1 approached. A primary matter under dispute between law and insurance was "fortuity." Insurance arguments were discussed in the previous chapter; this chapter deals with the position legal counsel was taking on the fortuity matter.

Sherin and Lodgen, LLP, were and are respected insurer defense attorneys headquartered in Boston. I was sent a white paper written by Benjamin Love, Esq. of Dallas-Fort Worth and edited by Barbara O'Donnell at Sherin and Lodgen. Love clearly summarized the possible coverage assertions for Y2K expense claims, with both offense and defense positions. *Fortuity*, *expected/intended occurrences*, *foreseeable occurrences*, *contra view*, *subjective* and *objective standards*, and *known loss*, all terms used in insurance case law, were outlined for defense counsel. Four *trigger* theories were also outlined in the white paper draft, including a review of existing policy exclusions that might apply to Y2K claims as well as new exclusions specific to Y2K. While details of those legal defenses are beyond the scope of this book, Love's article for *CGL Reporter* was a valuable summary tool for Y2K claims defenses.

Love confirmed two important points: 1) "We have learned from environmental insurance litigation that disputes over who will pay for an ill suffered by society in general may result in payment of many insurance dollars to help 'society right a wrong.'" 2) "Given that virtually every court in the country has ruled that the duty to defend is broader than the duty to indemnify, insurers will be required to pay substantial defense costs even if there is not ultimate duty to indemnify." Love concluded with a proposal for the advantages of mediation for Y2K insurance dispute issues.[80]

It's worth noting here that a different rationale for asserting fortuity was proposed by California attorneys, who suggested in

March 1998 that sympathetic juries might be "[M]ore willing to consider a Y2K loss *accidental*—at least from the insured's standpoint—where the insured's liability stems from its dependence upon a third party for information, and thus on a computer system removed from the insured's own."[81]

In a mid-1998, 54-page publication by J&H Marsh & McLennan titled *Y2K and Your Insurance*, Marsh took the position that the issue of fortuity was not entirely decided, since no one could reasonably predict the impact of two-digit date fields, nor what central a place computers would have in commerce and government.[82] Clearly, fortuity was a hotly debated topic with respect to insurer's response to Y2K compliance costs.

I offer an amusing side note: A *Dilbert* cartoon on Y2K that ran September 17–19, 1996, was referenced by the American Bar Association as the early warning to the nation, given the popularity and widespread readership of *Dilbert*.[83] The *New York Times* "Investor" section republished the cartoon in the lead story regarding the stock market. That cartoon, mentioned in several publications, caused fortuity defenses to be then settled as moot, thanks to *Dilbert*.

A California attorney argued that asserting fortuity as a defense against coverage might come back to haunt an insurer when mounting its own defenses for internal Y2K glitches. Asserting fortuity, which caused claims processing problems, for example, cannot be used in defense of its own negligence suit(s) if the reverse is used in denying coverage![84]

Ultimately however, the ruling of the U.S. District Court in New Jersey said the insurers were not required to reimburse GTE for the money the company spent to repair and test its computers to address Y2K problems, arguing that computer software "flaws" are not covered, as reported in a September 2000 *ABA Journal* article coauthored by Robert Carter, Esq., of McKenna & Cuneo. Late notice was also a deciding factor in the court's decision, and it was reported that GTE's insurance broker had drafted their policy wording, which limited GTE's discovery options.[85]

The Xerox case against American Guarantee and Liability Insurance Company (AGL) was found for the insurer by a New York Supreme Court judge on December 20, 2000. The court concluded that Xerox has failed to give AGL "timely notice" as required by the insurance policy terms, thus denying the insurer the opportunity to investigate the claim. The basic requirement to bring in an insurer early in the process offers insurer supervision of both processes and expenses associated with an insured loss. In Xerox's case, a determination of possible remediation expense coverage versus normal computer upgrades was not afforded AGL. Xerox's assertion of coverage included approximately $138 million (reported variously in publications as $183 million) in Y2K remediation expenses already spent, filing claim notice in March 1999. Insurers customarily require notice be filed within 60 days of knowledge of a potential loss. Xerox was reported to have started Y2K work in 1993 and remediation in 1996. Its March 17, 1999, claim filing was the basis for the court's denial of coverage.[86] The decision on the Xerox case, however, left all other coverage arguments unresolved. All of these large cases were watched closely by insurers, attorneys, and their clients to see who would be paying for this massive Y2K remediation effort.

From the five sue and labor suits at the end of 1999, by early January 2001 there were eighteen pending Y2K sue and labor lawsuits.[87] "Fortuitous events" are a basic part of sue and labor clauses with the argument that fortuity was never a part of Year 2000 issues within the meaning of that coverage section.

In addition, standard exclusions of latent defect, faulty workmanship, intentional misconduct, and possibly obsolescence/wear and tear might be applicable. "Sudden and accidental" terms are commonly used in determining loss coverage on property policies.[88] These terms inevitably suggest no coverage for Y2K remediation; insurance policies as sold were not intended for this once-in-a-hundred-years issue. And once corrected, would not happen again in 2100, 2200, or beyond.

Attorney Peter Kelman advised *High-Tech Quarterly* readers on the matter of sue and labor. He, too, recommended clients discuss coverage with their brokers as well as their attorneys. Interestingly, Kelman also suggested asking corporate counsel whether filing a Y2K claim was an obligation required to shareholders.[89] The suggestion of that obligation was no small matter for corporate directors and officers.

Following the sue and labor cases in an April 14, 2000, *The Standard* article, the Insurance Information Institute Chief Economist, Robert P. Hartwig, reported that a handful of corporations and attorneys in the United States continue to look for ways to recover between $100 billion and $600 billion spent to rectify the problem: "During 1999, several dozen corporations and government entities decided to try to recover Y2K remediation expenses from their insurers through the obscure 'sue and labor' clause." Hartwig added that the millions of businesses that addressed Y2K without filing claims spoke volumes.[90]

In summary, insurance clauses requiring timely notice, a "fortuitous" event, "sudden and accidental" occurrences, as well as exclusions for latent defect, faulty workmanship, intentional misconduct, and possibly obsolescence/wear and tear were the bases for insurers denying reimbursement for Y2K remediation expenses. Thus, claims under "sue and labor" coverage were considered to be weak.

OTHER LEGAL CONCEPTS ABOUT "FINDING COVERAGE"

Business Insurance magazine carried a provocative article quoting Washington, DC, attorney Matthew L. Jacobs who encouraged insurance claims be presented when a government entity shuts down a business for non-Y2K compliance. The case cited a Georgia bank holding company, Putnam-Greene Financial Corp., which was issued a cease-and-desist order by the Federal Reserve Board, the Federal Deposit Insurance Corp., and the Georgia

Department of Banking when the bank's antiquated computers were feared to miscalculate interest payments after January 1. Jacobs maintained that the costs of FFIEC requirements for Y2K-compliance computer upgrades should be covered if shut down due to regulatory requirements.[91] As with sue and labor, fortuity (foreseeability) could be the primary argument denying coverage.

Francoise Gilbert, chair of the technology and intellectual property group at the Chicago law firm of Altheimer & Gray, noted in early and mid-1996 that software developers had filed lawsuits against independent service organizations (ISOs) providing maintenance and support for the developers' clients. Developers anxious to sell Y2K-compliant upgrades or entirely new software products acted to protect their interests from ISOs by recommending either fixes to current software or other vendors entirely. Apparently, no direct millennium upgrade failure suits had been filed. Suits at that point were market protection and revenue driven.

In *Mealey's Litigation Report: Insurance*, July 7, 1999, attorneys Scott Seaman, Esq. and Eileen King Bower, Esq. outlined their theory that there may be less litigation than might be expected by the $1 trillion expected by the bar. Their first item involved legislative statutes enacted to limit litigation, as well as the possibility of tort reform.[92] Seaman and Bower continued with obstacles to litigation: negative publicity, discovery procedures leading to unwanted public disclosure, damage to close business relationships, business judgment rule, the absence of "strict liability" standards, and interestingly, jury appeal of nonbodily injury claims. They added that noncoverage-related factors may lessen exposures for insureds, insurers, and reinsurers. The possibility of discovery requests of insurers' own Y2K readiness was also suggested.[93] These were very insightful thoughts as the stress of January 1, 2000 loomed.

My experience with this issue is that courts will seek funding from insurers when municipalities and government entities need money in extraordinary times. This is a different viewpoint than with corporate business entities.

DIRECTORS AND OFFICERS EXPOSURES TO LITIGATION

Issues related to corporate D&O liability associated with Year 2000 gained considerable attention as the new millennium neared. One of the best summaries describing those issues appeared in a December 1998 NewsEdge reprint from *Best's Review*, which cited the widely published damage estimates of $1 trillion in litigation, noting it would dwarf the $2 billion Dalkon Shield/IUD settlements. American International Group (AIG) and Chubb represented about half of the D&O liability insurance market. Again, the article referenced Armageddon when reporting Y2K problems facing corporate directors and officers. By late 1998, most large law firms had established Year 2000 practice sections, focusing particularly on insurance clauses and terms which might invoke defense and/or coverage for a Y2K-related claim. The well-known, and frankly feared, New York law firm Milberg, Weiss, Bershad, Hynes & Lerach was cited as having assembled an in-house team.[94] Failure to disclose corporate Y2K planning, preparedness, compliance, and costs, as well as those issues concerning protective legislation, were the focus of D&O liability and litigation. For all of the searching through every term, definition, and clause of general liability insurance policy language, ultimately the buck, that is, liability, rests with corporate management and board. D&O-directed litigation seemed to be a better bet than attempts for redefining general liability policy language.

The *Best's Review* article noted that D&O claims were the highest in the 1990s since studies began in 1974, with premiums dropping as new D&O insurers entered the market, providing competition to the major carriers, AIG and Chubb. In a Tillinghast survey, "31% of respondents had at least one claim filed against their directors and officers in the nine months prior to 1997."[95]

Litigation experts cautioned clients: "It is the legal obligation of Directors of U.S. companies to act as the 'stewards of the company's assets.' It is their job to plan and perform strategically. Thirty years ago, in the area of corporate systems, the failure of internal computers would have been an inconvenience. Today such a failure today can cripple a company."[96]

My own experience with D&O liability insurance policy renewals extending into the new millennium with my publicly traded clients was to assist clients with their detailed Year 2000 planning and milestone disclosures to their D&O underwriters. Those were unusually meeting-intensive years. I recall the discomfort of having to call my underwriter two days before the policies' renewal when one very large client's arrogant Y2K compliance manager consistently refused to provide Year 2000 remediation documents to me after numerous meetings for that purpose. I advised their insurer's underwriting manager to hold the renewal until I was assured that Y2K compliance was accomplished or well underway, notified the CFO, and ultimately got to see the compliance presentation that had been presented to their board. Eleventh-hour maneuvering became a predictable part of Year 2000 in nearly every respect.

In the October 25, 1999, issue of *National Underwriter,* Ara Trembly reported that shareholders might sue directors and officers if corporations don't aggressively seek insurance proceeds to cover remediation expenses as a failure of due diligence.[97] D&O lawsuits might also be filed by shareholders when stock values fell as a result of corporate Y2K problems being publicized, or for inflated pre–01/01/2000 compliance statements. More importantly, criminal liability could be asserted against directors and officers should mission-critical system failures from Year 2000 problems result in injury or death. It was also noted that criminal penalties for corporate securities law violations involving fraud or deceit involving Year 2000 could conceivably be brought. Understandable how nervous corporate management was becoming.

Testa, Hurwitz & Thibeault, LLP, established in 1973, was one of Boston's most prominent technology law firms. In their quarterly *Venture Update* publication of Summer 1998, the SEC and DOL Year 2000 Guidance was addressed. Of interest in the review of compliance guidelines, author David S. Godkin noted,

"In addition, as companies which are unable to meet the December 31, 1999 deadline consider a sale of the company as a possible solution to a Year 2000 problem, fund managers and directors of portfolio companies need to exercise increased diligence with respect to acquisitions."[98]

The contemplation of selling a non-Y2K compliant entity was rarely, if ever, addressed; a thought-provoking idea.

The difficulties that individual sectors were facing dominated media air time and print. Unfortunately, the greater story of collaboration never received much, if any, media attention. News accounts typically examined a single sector's Y2K issue without mention of the tremendous amount of cooperation going on behind the scenes. Responsibility for successful Y2K compliance was routinely directed anywhere and everywhere else, with risk avoidance appearing at the forefront of virtually every private and public sector.

It's understandable, then, that from an outside perspective, Y2K compliance success seemed far and out of reach.

Defense Research Institute, November 4 & 5, 1999, Boston, MA

by Nancy P. James, December 27, 1999

The Defense Research Institute, Inc., gathered for a specially scheduled program in Boston November 4 and 5 to take a hard look at defense of Y2K property and liability claims. Speaking before an overflow audience on a newly popular subject, law experts from all parts of the United States described how Y2K technical problems might shake out to cost insurers billions of dollars in either defense costs, covered judgments, or both. In fact, Diane L. Polscer, Gordon & Polscer, LLP, of Seattle, WA, stated that estimates of costs to insurers are between $15 and $35 billion for Y2K-related claims. However, many estimates are much higher, when defense costs are included as possibly covered.

Attorney Polscer offered some perspective of size, offering that Y2K will be second only to WWII in total calamity costs (WWII costs are stated to be $4.2 billion). In comparison, the Vietnam War is estimated to have cost $500 billion, and the 1989 California earthquake at $100 billion. The $600 billion Y2K remediation expense referenced by several speakers combined with the $1 to $1.5 trillion dollar defense costs exceeds the U.S. federal budget.

Speakers, one after another, reviewed the history of D&O and E&O, reflecting who might be vulnerable to litigation as a result of Y2K failures. D&O issues were raised such as the difficulties of assessing non-standard forms as well as carriers' reluctance to assert coverage, especially with exclusions such as "willful violation" of Y2K regulatory requirements as a defense against coverage.

Regarding E&O, computer consultants headed the list of professionals susceptible to Y2K E&O claims, with attorneys a close second. Notable also on the list were insurance brokers, with discussion centering around brokers allowing companies to place Y2K exclusions on 1999 renewal policies, as well as ultimately explaining to insureds those Y2K exclusions which have been placed on renewal policies. Professional standards of advising obligations will be scrutinized as Y2K injuries encourage litigation.

In summary, while many hours of legal research and analysis were gathered and shared, Year 2000 remains the biggest wild card practicing professionals in all disciplines have ever faced. Estimates of damages, business interruptions and losses vary with unprecedented magnitude. It was clear the defense bar is bracing for the Millennium, as are many professionals, insurance included.

6

LEGISLATION AND GOVERNMENT ON YEAR 2000 ISSUES

Numerous archived documents reference commentary on the projected effectiveness of Y2K-specific legislation. Below is a summary of some of the statutes enacted by Congress and the Commonwealth of Massachusetts, which serve as a chronicle of federal and state government's involvement in legislation, as well as insight into where protections were being directed. It is a summary of a complex regulatory climate, locally, statewide, and nationally.

With regard to the millennium, legislators and regulators uncharacteristically found they had to protect (i.e., regulate) not only those they oversaw, but themselves as well. This may have offered a singular opportunity for more sympathetic views of their constituencies. Think about that! But then, contrarily, legislators' own fears of Y2K noncompliance may have led to the enactment of harsher consequences. You be the judge.

In an *Insurance Times* front-page article by Penny Williams, interviewing me (March 3, 1998), I stated: "Courts may rewrite insurance contracts to find coverage." I also predicted that the federal government may step in to *help* the industry by possibly declaring Year 2000 a "virus" or a "natural disaster," either of which could affect the issue of fortuity, and invoke insurance coverage not anticipated at this time.[1] That particular act of Congress, specifically defining Year 2000 for economic or political purposes, was never realized, unlike the walk-back of President George W. Bush's "act of war" declaration on September 11, 2001, to "act of terrorism" one day later in order not to exclude *all* insurance proceeds for damage and lives lost in New York's World Trade Center twin towers tragedy.

UNITED STATES GOVERNMENT PREPARES FOR Y2K

In the mid-1990s, a U.S. House of Representatives committee estimated the total federal government cost for Year 2000 remediation at $30 *billion*. Interesting issues faced Washington as it acknowledged that there would be no legislative capacity for a "silver bullet" solution to Y2K code review and updating. Indeed, Washington could not solve this one by fiat! At the time, there were only 1,000 government software professionals who were identified as capable of Y2K remediation work, interestingly summarized as a single programmer needing 470 years for Y2K project completion.

Further analysis of the magnitude of the issue noted that only 16 percent of all computer-based projects are actually completed on time.[2] Yet 01/01/2000 was not a date available for recalculation!

In July 1996, the Subcommittee on Government, Management, Information, and Technology offered a Y2K readiness scorecard for each of the twenty-four federal agencies. Fourteen of the twenty-four agencies received a D or an F. Justice, NASA, and Veterans Affairs were among the ten rated who were given a D grade; FEMA, Labor, Energy, and Transportation made up the four receiving a grade of F. Considering how important these entities are to the functioning of American life, this was clearly serious. A September 1996 joint hearing with the House's Science and

Technology Subcommittee quoted its chair, Constance Morella, as concluding, "The consequences are pretty grave." Among the most at risk for systems failures included transportation, air traffic control, and defense systems. Furthermore, at that September 1996 hearing two cabinet secretaries were identified as having no awareness of the Year 2000 computer issue at all! Improvements in the next two years were given mixed reviews; in August 1998, California Republican Stephen Horn, chair of the subcommittee, issued a scathing report card of Ds or Fs for most large government agencies' Y2K readiness.[3] John Koskinen, chairman of President Clinton's Year 2000 Council (known as the country's "Year 2000 Czar"), replied that readiness was more of a C+ or B. While it might not be hard to imagine how "ready" might realistically be measured at that time, caution is always the safer course, optimism being judged more harshly in 20/20 hindsight.

Analogies were common for the purpose of graphically describing to the public complexities that were genuinely and understandably too arcane to be understood by the public at large. In mid-1996, Sally Katzen, administrator of Information and Regulatory Affairs in the Office of Management and Budget, suggested that ongoing fixes after 01/01/2000 "invoked the analogy of rebuilding a rocket ship while it is on its way to the moon."[4]

The Social Security Administration (SSA) had 35 million lines of software code leading up to January 1, 2000, according to Associate Commissioner Kathleen Adams.[5] Thirty-five million *anything* is a magnitude difficult to grasp, let alone prioritize. And yet, the SSA did! The SSA, having experienced a software issue as early as 1989, then began its Y2K investigation and remediation, estimating a $30 million cost and 300 staff years for successful completion. The SSA was one of only four agencies receiving an A grade for Y2K readiness in a September 1996 federal review. President Bill Clinton announced on December 28, 1998, "The Social Security system is now 100 percent compliant with our standards and safeguards for the Year 2000."[6]

Year 2000 studies were undertaken by the 105th U.S. Congress, including a Special Committee of the Year 2000 Technology Problem chaired by Senator Robert F. Bennett and Senator Christopher J. Dodd. This resulted in their early 1999 report, "Investigating the Impact of the Year 2000 Problem." The executive summary addressed the fears of litigation and loss of national competitive advantage, proposing legislation outlined below. Gartner Group studies were widely cited to focus on the importance of comprehensive readiness to protect the country. The title page offered a quote by Henry Kissinger: "*Competing pressures tempt one to believe that an issue deferred is a problem avoided: more often it is a crisis invited.*"[7] Bennett and Dodd's committee held approximately thirty hearings in the year running up to the new millennium.[8]

Senator Bennett issued a press release on June 12, 1998, with a troubling assessment of the nation's power grid. Based on his committee's survey, Bennett concluded, "I am genuinely concerned about the very real prospects of power shortages as a consequence of the millennial date change." Only 20 percent of those surveyed had even completed a Y2K assessment, with *none* reporting comfort with their vendor and supplier readiness. Reports of seriously failed Y2K tests were described with conclusions of the dangerous consequences should January 1, 2000, be a failure.[9]

Testifying before the U.S. Senate Select Committee on the Year 2000 Technology Problem on July 6, 1998, was John Westergaard, publisher of Westergaard Online Systems, Inc., internet magazine. He had some sharply unfavorable comments in his testimony, criticizing the Department of Defense, Army, Navy, and Air Force, the Y2K response of Japan and China, as well as the *Wall Street Journal* for "abysmal" Year 2000 coverage. Calling Y2K an "electronic bubonic" plague, Westergaard took credit for early advising Senator Patrick Moynihan to request of President Clinton a "Y2K Czar;" Clinton ultimately named John Koskinen. Westergaard testified that Koskinen "occupies the fourth most important position in America today after the President, Fed Chairman Greenspan, and Treasury Secretary Rubin."[10]

One year later, the President's Council on Y2K Conversion in its second quarterly 1999 report summarized the status of varied readiness in extensive industry and financial sectors: Federal mission-critical systems would be compliant; national infrastructure readiness was anticipated; international Y2K compliance efforts were behind and at considerable risk. Small businesses, local governments, and healthcare providers varied greatly in readiness, posing risk to dependent entities. With a scant six weeks to go before the inflexible deadline of January 1, 2000 arrived, the federal government issued an optimistic report card with many As for agencies' readiness. Perhaps not surprisingly, many agencies were left off of that report card, specifically Medicare and Medicaid. Reports from Congress indicated that many programs were still unprepared.

This section would not be complete without mentioning complexities compounded by the Internal Revenue Service. Reporting of Year 2000 expenses could be scattered among difficulties of classification; replacement costs, capitalization, amortization, consulting, hardware, software, or labor costs. Accountants began advising their clients of proper documentation of all Y2K-related expenses.[11]

It was my prediction at the time that the IRS would be very unforgiving for late filings and impose fines as usual, notwithstanding taxpayers' Y2K complexities and processing delays. In 20/20 hindsight, however, that was not reported to have happened.

In conclusion, the above is just a small sample of the focus of national government experts and officials as year 2000 readiness was being analyzed, rated, legislated, and undertaken.

CONGRESS ENACTS YEAR 2000 LEGISLATION

A number of bills were passed by the 105th Congress to provide preemptive defense against Y2K litigation as a result of predictions that litigation expenses for Year 2000 issues could reach $1 trillion to $1.5 trillion. In addition, it is important to note that then and now, government entities—local, state, and federal—are careful to consistently hold themselves harmless from liability. Legislation has been carefully and exhaustively documented and analyzed in many subsequent publications, so what I outline below serves to point out the focus and priorities of our elected representatives.

H.R. 4455, Year 2000 Readiness Disclosure Act was legislation introduced in September 1998 by Congressman David Dreier (R-California) and Congresswoman Anna Eshoo (D-California) to encourage companies to share Year 2000 preparations without fear of litigation.[12] The Senate version, Year 2000 Information & Readiness Disclosure Act (IRDA) Senate bill S.2392 of October 19, 1998, protects Y2K statements, readiness disclosures, and antitrust safe harbor, except in cases of direct intent, reckless disregard, and bad faith. Congress called the bill the Good Samaritan Law. The Senate version was a compromise measure by Senators Orrin Hatch and Patrick Leahy, designed to minimize Year 2000 disclosures from being used in court proceedings, which ultimately led up to enactment of what would be known as "the Act" with President Clinton's signature.

The powerful U.S. Chamber of Commerce issued a press release hailing the compromise bill, with Lawrence Kraus, president of the Chamber's Institute for Legal Reform, stating, "Companies are working as quickly as they can to solve the Y2K problem. But they are crippled by a reluctance to share information about potential hardware and software glitches because of trigger-happy trial lawyers seeking to exploit the situation."[13] The American Insurance Association (AIA) was reported by the *Insurance Times* in September 1998 as very much in support of the legislation. Melissa Shelk, AIA assistant vice president of federal affairs, stated, "No one will benefit if widespread disruptions occur due to Y2K problems."[14]

In the same *Insurance Times* issue, a lengthy editorial expressed concern over the lack of time left to enact meaningful protective Y2K legislation. Detailing apprehension over solutions to the insurability issues of Year 2000 compliance, editors noted how much more difficult it would become if courts were involved: "Courts are more concerned with affixing blame than with finding answers." The editorial concluded by referencing my "modest" proposal of a Millennium Mediation Council, calling it a "vehicle of hope for businesses, insurers, consumers, and lawmakers who desire to find a rational way to survive the Year 2000 and succeed in the new century."[15]

During Congressional deliberations on the bill, a brief article in the *New York Times* on September 21, 1998, reported, "Senator Joseph R. Biden Jr., Democrat from Delaware, stated during a heated Judiciary Committee debate 'What this bill does is lower the standard of care for a supplier.'" The article noted that the compromise efforts were being assailed by trial lawyers fearing that computer and chip makers would be granted immunity from liability suits. Opponents used terms like "catastrophic problems" and "shirking responsibilities" to refer to corporate preparedness measures, while proponents supported disclosure and antitrust protections for information sharing.[16]

A noteworthy article in the *New York Times* in early December 1998 reported that there was concern by enterprises about their Y2K disclosures, resulting in greatly increased requests for legal assistance. The bill's complexity compounded corporate confusion, including the forty-five-day window following enactment to bring past corporate Y2K statements under its protection. Clinton's Y2K head Koskinen explained that the bill's critical function was to provide protection for the sharing of more technical information, not exclusively readiness statements. Koskinen encouraged trade group clearinghouses for Y2K information.[17] But remember, corporate concerns primarily involved the limitation of massive litigation expenses.

Understandably, neither IRDA nor "the Act" were intended to protect "an allegedly false, inaccurate or misleading Year 2000 Statement." Disclosure statements under the Act had both an expiration date (December 3, 1998) and retroactive date (January 1, 1996), and were dependent upon parties not having "already relied upon the statement prior to receiving notice."[18] Indeed, confusing enough to need legal counsel. According to Chicago attorney José Isasi, the IRDA was not thought to be a protection for computer hardware or software companies because their alleged "failure" (referring to the two-digit date field) would have taken place prior to the Act's January 1, 1996, retroactive date. Ultimately, the Act's protection would be offered to Y2K compliance firms, corporate directors and officers, and associates such as accounting firms.[19]

On February 10, 1999, Senate Commerce Committee Chairman John McCain (R-Arizona), held the first full committee hearing to limit litigation and civil actions for Y2K failures. The committee's ranking Democrat, Senator Ernest Hollings, opposed such legislative actions, stating he would try to "kill [the Act]" because "it is a gimmick."[20]

The U.S. Chamber of Commerce had recommended legislation strictly limiting Year 2000 litigation far exceeding the October 1998 IRDA. The Chamber's proposal included special Year 2000 courts, limits on joint and several liability, and limits on contingency fees and arrangements with lawyers. On June 15, 1999, the U.S. Senate overwhelmingly passed a measure to restrict millennium-bug liability lawsuits with a 62–37 vote; the U.S. House passed it 236–190. The Chamber of Commerce hailed the measure, which included a cap on damages at $250,000 for small businesses. President Clinton pledged to veto that bill, but ultimately supported the compromise "Y2K Act" restricting class-action lawsuits and capping damages.[21] H.R. 775, as passed, makes no mention of insurers.

H.R.775 YEAR 2000 READINESS AND RESPONSIBILITY ACT ("Y2K ACT")

On July 21, 1999, President Bill Clinton signed into law "Y2K Act," a bill similar to Senator McCain's Senate Y2K S.96 bill to encourage correction rather than blame. Ultimately the Y2K Act established that Y2K problems from January 1, 1999 to January 1, 2003 were subject to regulation to encourage Y2K problem resolution before seeking court action. The Act restricted Y2K-related actions and damages, upheld the enforceability of contract warranty disclaimers, and required written notice by plaintiffs to each potential defendant in advance of filing a Y2K lawsuit and a 90-day stay period for resolution prior to proceeding with the suit. However, the Act was largely not applicable to securities-related class-action suits.

A *New York Times* editorial on July 3, 1999, regarded Clinton's agreement to the Y2K Act's compromise an "ill-advised turnabout," and stated that the legislation would reduce the rights of legitimate claimants, lessen industry incentives to correct Year 2000 problems, unwisely suspend joint and several liability, and limit financial liability; all a departure from tort law principles. The editorial concluded with a reference to the administration stating that the Y2K Act was not a precedent for further tort reform.[22]

Nevertheless, Senator McCain's S.96 bill was supported by powerful business groups joining the U.S. Chamber; the National Association of Manufacturers, and the American Insurance Association, among many others.

Vocal critics of government addressing the issue at all were specific in their complaints. "In fact, the U.S. government is abdicating any kind of leadership role in this crisis. Government agencies are among the worst prepared in the country," wrote Edmund deJesus in *Byte* magazine in July 1998.[23]

My "Modest Year 2000 Proposal: *A Millennium Mediation Council*" guest editorial was also run in the mid-September 1998 *Insurance Times*.[24] My article and proposal did not reference the urgency of legislation, but offered a mediation solution, pointing toward the staggering costs of litigation coupled with complexities of insurance contract interpretation for this new peril. I have described my position more appropriately in chapter 7.

H.R. 3230, the Defense Authorization Act (September 1996) required the Department of Defense to evaluate the impact of the Y2K problem and report to the House National Security and Senate Armed Services Committee by January 1997.

In Executive Order 13073, President Clinton created the President's Council on Year 2000 Conversion (February 4, 1998), to address the Y2K challenge, nationally and internationally. The Council, made up of senior government officials for twenty-five working groups, coordinated the government's overall Y2K activities. The Council included key infrastructure areas by forming a Senior Advisors Group (SAG) in January 1999, consisting of more than twenty Fortune 500 company CEOs and heads of major U.S. public sector organizations. Executive Order 13073 committed to federal readiness, "cooperation with state, local and tribal governments to address the Y2K problem," cooperation with private sectors, banking and financial, telecom, public health, transportation, and electric power systems.[25]

The Securities and Exchange Commission (SEC) in October 1997 had established Y2K readiness disclosure guidelines for publicly traded companies, updating it in 1998 to include Year 2000 remediation costs and likelihood of "material financial impact" on the enterprise. Further directives added requirements to disclose worst-case scenarios and costs, as well as the existence of any contingency plans. The SEC, concerned that their earlier guidelines produced only vague and limited Year 2000 compliance reporting, provided more specific Year 2000 guidelines for the roughly 10,000 publicly traded U.S. companies. A letter issued to over 9,000 chief executives by SEC chairman Arthur Levitt stressed the importance of investor confidence in providing meaningful disclosures on Y2K preparedness and contingency planning.

By mid-1998, Congress was expressing impatience with the SEC for more robust guidelines, which had not yet been issued. At a hearing, Senator Robert Bennett, chair of the Special Committee on the Year 2000 Technology Problem, accused the SEC of dragging its feet, stating that corporate Y2K disclosers were "abysmal."[26]

As the new millennium loomed, the SEC in a 5–0 vote set a target date of August 31, 1999, for brokerage firms to be Year 2000 compliant, with November 15 as a final deadline. As of December 1, 1999, the SEC would shut down noncompliant firms through court orders. The *Boston Globe* reported that by the end of July 1999, it was estimated that only 1 percent of transfer agents were not Y2K ready. Boston law firm Robinson & Cole, LLP, noted that during 1998 and 1999, ninety-six broker-dealers had been charged by the SEC for failing to report their Year 2000 costs and plans as required. Disruption of shareholder transactions and dividend payments was the SEC's understandable concern.[27]

The Federal Financial Institutions Examination Council (FFIEC, which oversees and coordinates the banking sector's FDIC, FRB, NCUA, OCC, and CFBP) issued guidelines for financial sector D&O compliance, disclosure, and duties. In addition, federal agencies overseeing health care, small business, and food and drugs provided information and resources for Y2K.[28]

Leading up to what were the specific unknown legislative consequences of problems arising from the "Y2K Bug," the media was filled with experts' predictions. As reported by Alex Maurice in the *National Underwriter* August 30, 1999, issue, attorney Doug Ey of Smith, Helms, Mulliss & Moore, LLP (Charlotte, NC), commented that "Congress has effectively created a new legal system for one particular type of claim." Ey told Maurice that he feared that "everything is going to be fought out. The constitutionality [of the law] is going to be questioned." Maurice reported Ey noting an example of a North Carolina law requiring mandatory mediation before any other action is undertaken, thus questioning the 90-day window for Y2K error correction. "The constitutional issues are going to be really interesting," Ey stated again. "Some people say this Congress has exceeded its authority on this."[29] In early 1998, North Carolina was reported to be interested in encouraging states to band together for class-action lawsuits against computer companies, specifically to recoup their roughly $132 million remediation expenses.[30]

Maurice in the same article went on to reference Roger Andrews, first vice president of the Risk and Insurance Management Society in New York, as more optimistic about H.R. 775 (Y2K Act) to "limit potential litigation." Eric Goldberg, senior counsel for the American Insurance Association in Washington, DC, was quoted as stating "the centerpiece of the law is to preserve contract terms. Most of these [Y2K] suits allege breach of contract or warrants."[31]

Some Year 2000 authorities saw the Y2K Act as a possible detriment to responsible Year 2000 compliance, as the bill had provisions limiting damages, a thirty-day correction period which could have an additional sixty days attached, along with complications for duties imposed on the injured party. A thirty- to ninety-day waiting period for court hearings was also part of the bill, which critics suggested would not help injured businesses in speedy resolution to their issues. Some concluded that the legislation would indeed lead to *higher* Y2K legal costs for businesses. Other opponents of Y2K protective legislation argued that consumers would be hurt by noncompliant products continuing to be manufactured and sold.

From the technology perspective, Edmund deJesus, critical of governments' enacting limitation of liabilities, stated that "this helps most those who do the least," going on to note Nevada and Washington state "shrewdly passed legislation protecting the states themselves from such litigation."[32] Technology attorney Jeff Jennett noted that "Georgia, Virginia, and Hawaii had passed laws to shield themselves against possible lawsuits if their computers failed." Jennett offered an example of lawsuits being barred if late benefit checks due to Y2K glitches delayed critical medicine purchases with life-threatening outcomes.[33]

There was rigorous debate in the Senate, with Republicans supporting legislation prohibiting or limiting Y2K litigation, thought at the time to be so potentially massive to cause an economic recession. Senators John McCain (R-Arizona), Robert Bennett (R-Utah, chairman of the Senate Special Committee on Year 2000

Technology Problem), and Conrad Burns (R-Montana) were leading supporters of such legislation, while Senate Democrats like Ernest Hollings (D-South Carolina), Ron Wyden (D-Oregon), and John Rockefeller IV (D-West Virginia) were opposed. Senator Christopher Dodd (D-Connecticut), chairman and ranking Democrat of the Senate Special Committee on the Year 2000 Problem, assured his Democrat colleagues that the problem was real, and needed legislative protections against litigation. Senator Bennett was quoted as stating, "Fear of litigation is so great that it threatens to eclipse the goal of remediation," adding that the monetary stakes could be three times the nation's insurance pool and lawsuits three times the U.S. court system's current 1999 caseload. National debate at the time over Y2K legislation described plaintiffs' attorneys as "salivating" over potential Y2K litigation, and "mother lode" comparisons with awards from tobacco litigation.[34]

Infuriated Democrats openly accused the White House of selling them out while Republicans applauded the bill's passage. Senator Dodd represented Hartford, the nation's capital of insurance, and was anxious to protect its interests. President Clinton's political complexities surrounding the Y2K Act were many; if he vetoed the bill, placating Democrats, he would hurt Al Gore's presidential aspirations by alienating Silicon Valley. Conversely, trial attorneys were historically big contributors to Democrats, while business interests were substantial donors to Republicans.[35] Year 2000 had its own set of political troubles, indeed.

On top of that, remember that President Clinton had been impeached in December 1998 for lying under oath and obstruction of justice in the Paula Jones and Monica Lewinsky scandals, and had just been acquitted by the Senate in February 1999. With the Y2K Act, Congress was attempting to get back to focusing on urgent government matters.

I need to note that the Senate voted 99 to 0 on March 2, 1999, to authorize governmental loans to small businesses to fix computer problems.[36] What an extraordinary contrast then to Washington more than twenty years later. Can we look back in envy—and in our natural expectation not to expect otherwise—at the "good old days" when our Congress worked for our better (if not best) interests?

The divisiveness in our current political arena hinders the type of concerted cooperative effort that was experienced during Y2K compliance. This was particularly evident with the COVID-19 global pandemic. Indeed, Y2K should be considered a roadmap for crisis management. Uniformity of purpose in every sector, affecting every sector, contributed to a singular response to remediate the matter and keep civilization running.

Senator Bennett wrote an article published in the *Wall Street Journal* in September 1998 describing the importance of attention to Y2K, stressing that the problem was global, not just national or local. The article mentioned Bennett visiting a power plant in California with a well-informed CIO who reserved 15 percent for remediation and 70 percent for testing. His article concluded with a reference to both Paul Revere and Chicken Little, alerting that the problem is coming, but the sky has not yet fallen.[37] A New York court determined that insurance litigation was not specifically covered by the Y2K Act, but only "damage directly resulting from Y2K failure."

So what happened? The U.S. General Accounting Office (GAO) published a report on Year 2000 litigation on September 25, 2000 finding only a hundred Year 2000 state and federal lawsuits, mostly filed against company hardware or software vendors for the cost of upgrading existing systems. Senator Patrick Leahy asked the GAO to study the Y2K Act and was quoted as saying the GAO report "confirms that in the courts, the Y2k [*sic*] bill was mostly used by big companies to delay or sidetrack relief to consumers," calling it a "lesson for special legal protections."[38] An earlier September 1, 1998 GAO publication of testimony before Congress described the stark realities to government and business enterprises if Year 2000 software compliance was not undertaken successfully. The thirteen-page report was heavily footnoted with a lengthy bibliography following.[39]

AT THE STATE AND LOCAL LEVELS

The latest census data leading up to January 1, 2000, was the 1992 census, identifying over 84,000 local government entities in the United States. Washington's Bureau of National Affairs published a report in mid-1998 outlining some of the legal issues facing local governments, whose wide-ranging responsibilities include police, fire, schools, transit systems, airports, hospitals, water and sewer facilities, courts, and criminal justice systems. Legislated operational constraints including civil service statutes, labor agreement restrictions, as well as public bidding requirements, creating significant roadblocks to finding the most talented staff for Y2K compliance work.

While such constraints slowed remediation progress on critical municipal infrastructures, the doctrine of sovereign immunity designed to protect state and local governments from litigation had been largely replaced by tort liability statutes in every state, allowing suits to be brought by citizens harmed by negligent municipal employees.[40] We can conclude that situation to be an unenviable position for states, cities, and towns competing for Y2K software talent.

As states labored on their individual systems for Year 2000 upgrade and compliance, officials were asking federal authorities for financial assistance with upgrades to systems that interact with federal programs. State officials also expressed concerns that late-issued federal directives might counter or slow progress already made. In September 1996, with only thirty-nine months to the deadline, 75 percent of states reported they were in the planning stage for Y2K conversions, according to Daniel Houlihan of the National Association of State Information Resource Executives. He noted the unlikelihood that individual state solution initiatives would be compatible with one another and that no common standard had been set by federal agencies.[41] Those states undertaking Y2K in 1996 had cost estimates varying from $300,000 to $97,000,000. By February 1998, the estimated cost for state governments to fix the bug ranged from $1.5 million to $200 million. This is offered to describe the enormous variance/delta in expected costs of Year 2000 remediation among the states. The unknowns were staggering.

Network to network, government data transmission was still seen as being problematic, even with newer Y2K-compliant systems. Government systems communicate regularly with one another; with federal, state and local computers; as well as with the private sector, such as banks. Harris Miller, president of the Information Technology Association of America, stated, "The Year 2000 software conversion is arguably the largest and most complex global information management challenge society has ever faced." Glen Mackie, information systems manager for the Nebraska Department of Motor Vehicles, said, "If you are just storing and retrieving information, it's a cakewalk. If you're calculating penalties and interest, it becomes quite a religious experience."[42]

Steve Kolodney, chairman of the National Association of State Information Resource Executives (NASIRE), in 1996 noted that politics as usual needed to be considered in this once-in-a-lifetime event. Legislators were being asked to put aside local patronage to allocate significant funds for a less-than-high-visibility project. He colorfully expressed, "That's a low flying turkey that doesn't take a high-powered rifle to shoot down."[43] Back-office expenses of significant magnitude were a hard sell at best; an unenviable position for legislators.

COMMONWEALTH OF MASSACHUSETTS LEGISLATION

I will focus briefly on the Commonwealth of Massachusetts efforts surrounding Year 2000 issues. Massachusetts is known as a very blue, very liberal state with progressives in office both state and local. Leading up to the new millennium, I was active locally, offering two forums in Concord regarding infrastructure, law enforcement, and civil defense readiness. The first forum was in early 1999. During those sessions, I facilitated discussions among Concord residents, town officials, and business owners regarding the specifics of each compliance effort. I describe more fully in chapter 10 the details of local citizen concerns and the responses of town administrators and government officials.

Interaction in Massachusetts followed, where State Senator Susan Fargo was Senate chair of the Joint Committee on Energy and vice-chair of the Joint Committee on Science and Technology. Following attendance at my January 22, 1999, forum, Senator Fargo hosted an assembly the next month including local officials. She stated, "My concern is that people aren't talking to their (town) finance committees," citing budgetary issues regarding compliance costs. Fargo's forum included state Information Technology Division Counsel Ray Campbell, who cautioned municipalities, "It is really important that you don't just give this to your (computer) people. It would be a mistake to believe there are no liabilities."[44] As mentioned earlier, municipal immunity to citizen lawsuits had eroded steadily over time, leaving public officials facing "should have known" grievances from residents. Campbell's cautions were well founded. (Senator Fargo was a panel member of the wrap-up forum I presented to Concord in September 1999.)

Massachusetts had Year 2000 bills prior to the millennium turn: Senate Bills 503 and 504, protecting hospitals and healthcare systems from liability for both preparedness and Y2K failures; House Bill 2561, providing governmental immunity; and Senate Bill 834, limiting the commonwealth's liability for "Year 2000 problems."[45] The Massachusetts Department of Revenue issued a "Bulletin to Local Officials" dated October 1998 outlining "Year 2000 Compliance and Municipal Liability." The white paper included an organizational grid of municipal departments with Year 2000 affected segments detailing Y2K remediation needs. Referencing the Massachusetts Tort Claims Act of 1978, which limited liability damages to $100,000, the white paper cautioned that suits from numerous plaintiffs could be a catastrophic multiplier.[46] The Commonwealth of Massachusetts Department of Economic

Development mailed businesses a "Conversion 2000: Y2K" kit in November 1999. The kit contained a disc, a letter from MDED Director Carolyn E. Boviard, and several glossy pages outlining Year 2000 assessment and planning.[47]

In an interesting lead article in early 1998 in *Mass High Tech* on state government readiness and cost allocations, Charlie Baker, then secretary of the Executive Office for Administration and Finance, reported that the commonwealth had been upgrading computers for the past five years, reducing Y2K impact. The article goes on to report that the state's Information Technology Department (ITD) authority had been expanded for Y2K, creating a separation of powers issue between the ITD and the trial courts, which were accustomed to requesting information from the legislature, not the other way around.[48]

STATES DIVISIONS OF INSURANCE LEGISLATE Y2K INSURANCE MATTERS

Even while municipal, state, and federal governments were struggling with their own Y2K compliance issues, state insurance divisions were legislating what insurers could and could not do with policy renewals as Year 2000 approached. Since insurance is regulated independently by the fifty states since the McCarran–Ferguson Act of 1945, there has remained a cumbersome process of state-by-state compliance. However, many insurance policy forms are developed by a national body, Insurance Services Office, Inc. (ISO), which are then presented to the states' divisions of insurance for acceptance. As a consequence, there tends to be a uniformity within state insurance policy provisions nationally, which is primarily necessitated by business enterprises with operations throughout the country requiring consistent insurance coverage terms, state by state.

ISO developed Year 2000 policy exclusions, which were greeted with resistance from both the public sector as well as agents' associations. The Independent Insurance Agents of America (IIAA), one of the largest trade associations, initially opposed the exclusions but reversed their opposition in mid-1998, taking a neutral position. Reported in *The Standard*, IIAA in late 1997 "urged its state affiliates to oppose approval of ISO's endorsements, citing concerns that insurers would use them over-zealously in order to minimize their liability."[49] In my opinion, this is another example of "people unclear on the concept." By mid-1998, the magnitude of global remediation costs, litigation, and potential claims expenses clearly called for an examination of insurance coverage terms. The fact that IIAA was opposing exclusionary endorsements based on "indiscriminate use" and not based on the most pressing issue, which argued that coverage existed in the unendorsed policy, only shows how little IIAA officials understood the issue.

The state of Illinois, for example, issued a directive on April 1, 1998, called the "Illinois Compromise," stating what coverage forms could allow Year 2000 exclusionary endorsements. Personal lines policies (home and auto) would be prohibited from any Y2K exclusions, while commercial lines (business insurance) policies were given specific directives for Y2K endorsements, nonrenewal notifications, and deductible changes.[50] Richard D. Rogers, deputy director, Consumer Division of the Illinois Department of Insurance, in an interview with *Independent Agent* magazine stated that the directive served the interests of all parties, most particularly those of the consumer: "I believe in the marketplace. Departments of insurance must provide the industry latitude to operate freely while regulating. The corollary to this is that the industry must act responsibly and not abuse this privilege." Rogers went on to state that the Y2K issue would be carefully monitored for "inappropriate Y2K-related activities," and rigidly enforce all existing regulations as well as "sponsor new legislation to control insurance industry misconduct, if necessary. If companies abuse this privilege they will have the wrath of the DOI to deal with. And I don't think that is in their best interest."[51]

The California Assembly Judiciary Committee, on the other hand, in an effort to not discourage software companies from doing business in the state, proposed a bill that would prohibit all but actual damages from being awarded in Y2K suits. California, reasonably expecting a blizzard of lawsuits and attempting to avoid a technology exodus, filed the bill with the support of the powerful California Chamber of Commerce. As expected, attorneys objected, arguing that the bill was flawed; it would not serve to support high-technology sector growth, and should not serve to discourage corporate directors from their Y2K due-diligence compliance discloses and efforts.[52]

In an *American Banker* article reprinted by NewsEdge in early 1998, the Florida Insurance Department's Steve Roddenberry delayed approving ISO's Y2K exclusions, stating, "We are reluctant to give insurance companies carte blanche." The article went on to quote Steve Liner, regional manager, Alliance of American Insurers, referencing the Florida Year 2000 exclusions and coverage issue: "[S]everal recent court decisions have said insurance companies have no obligation to tell their clients that their insurance coverage (for Y2K) was inadequate, unless their client had specifically requested it." David Turner, vice president, Independent Insurance Agents of America, cited the cost of current Y2K insurance policies as cost prohibitive, saying, "As companies see opportunity for market share, we'll see more companies offering (Y2K) policies at a more reasonable level."[53]

I am unaware of any such Y2K court cases, as Liner referenced, with decades of case law rulings acknowledging insurers' and agents' duties to advise. In addition, David Turner's prediction of more affordable Year 2000 insurance coverage did not materialize. Reports such as this point out the scope of professional opinion based more on hope than fact.

By mid-1999, Florida's governor, Jeb Bush, signed legislation ultimately protecting both businesses and insurance companies from frivolous lawsuits. From protections for good-faith endeavors, proof of reasonable mitigation efforts, dollar damage limits, and arbitration and mediation provisions, the legislation was supported throughout the business and insurance communities. Interestingly, the original legislation had required casualty/property insurers to offer $1 million in Y2K coverage to enterprises specializing in Y2K compliance, effectively shifting liability from Florida businesses to "solution experts."[54] The staggering consequences of such legislation would have been mandating liability to the insured party, under duress—those Y2K computer software consulting teams, irrespective of their clients' negligence. While this was not an unheard-of "social benefit" strategy, common sense ultimately dictated a more reasonable and ethical distribution of liability. In addition, common sense suggests that insurers would immediately determine Florida software consultants to be an uninsurable class; ergo, a disappearing class. Naturally insurers balked, but more importantly, those unintended consequences would have seen casualty insurers fleeing Florida had those provisions not been dropped.

My own local regulators at the time, the Massachusetts Division of Insurance, released guidelines to insurers in October 1998. Y2K policy exclusions were prohibited in all but a few defined commercial policies, with nonrenewal notice provisions strictly enforced.[55] This is evidenced from the directives forwarded from trade associations like MAIA (Massachusetts Association of Insurance Agents) that the Division of Insurance was acting more on political motivations than informational.

While all that was going on, state by state, the NAII (National Association of Independent Insurers) trade association issued a press release on July 14, 1998, requesting that all states adopt a moratorium on new insurance division mandates from July 1, 1999 to June 30, 2000. A normal part of the arcane business of insurance, which is regulated independently by each of the fifty states, are annual January 1 rules and regulations, adjustments, and so on to current regulations, issued with volumes of supporting documents. Postponing those customary changes until April 1 would assist Year 2000 compliance focus internally and for clients.

The moratorium on regulatory changes, overwhelmingly supported by the National Association of Insurance Commissioners, was urged in order to give insurers "every opportunity to maintain the appropriate level of service to policyholders through this transition period."[56]

Here we have seen that not only did corporate America have a monumental struggle to identify, analyze, and modify vast quantities of computer code, but the need to comply with ever-increasing and ever-changing Year 2000 legislation and statutes added to those burdens. Probably no other time in American history had so many obligations been placed at the same time on the country's commerce. And remember, this was one deadline that purportedly could not be moved (though it was, with *windowing*!). Pausing to absorb the chronicle of these pressures, the success of Y2K remediation seems almost miraculous!

7

ISSUES, PROBLEMS, AND LITIGATION SOLUTIONS

By 1998, it was clear to our country's government and businesses that this Year 2000 thing needed attention. Attention and some deflection. I continuously received Y2K checklists from both my business partners as well as from many collegial Boston associates. While many of the checklists were well thought out and helpful to some extent, many more were comprised of a half-page of questions with compliance inquiries of vast and useless scope. One could not help but pause and ask whom the intended protection was for, and conclude that it certainly wasn't intended to help my clients, or me as their business partner. In my opinion, such correspondence was the customary and usual efforts of corporate attorneys to protect their clients by transferring risk. A more jaded explanation might recall the bar's trillion-dollar litigation expectation to be establishing targets.

Attorneys Stephen M. Honig and Lawrence R. Kulig in *Mass High Tech* addressed the issue of "compliance letters" in May 1998, offering that there is no accepted definition of "compliant." Noting both legal liability issues of careless replies as well as no actual legal obligation to reply to compliance requests, the authors noted: "In fact, some legal commentators believe nonresponse may even constitute 'anticipatory breach' of contract, inviting litigation." Advice included designating a single person for Year 2000 responses to help to assure uniformity of information, as well as factual website reporting that "does not insure parties against loss by reason of your Year 2000 noncompliances."[1] As with regulation and legislation, such efforts primarily served more as a distraction from the real goal of Y2K compliance efforts.

In addition, as both corporate and government entities embarked on Year 2000 compliance—some earlier than others—the issue of reported patronage arose. An early 1999 *Boston Globe* article reported that the Massachusetts Bay Transportation Authority had hired service contractors with political connections to oversee Year 2000 compliance and were over budget and behind schedule. Former MBTA officials reportedly said that no Y2K remediation contracts were awarded under competitive bid processes; qualifications evaluation were exclusively used to award work.[2] So we can see from that perspective, Y2K was no different than government as usual, likely occurring throughout the country.

ROGER SUDBURY

An account of my meeting Roger Sudbury belongs in this chapter to point out the challenges of successful crisis management.

In early May 1998, I was fortunate to meet Roger W. Sudbury of the Massachusetts Institute of Technology Lincoln Labs. Though we didn't discuss Y2K directly, his work concerned similar date testing. In the course of our conversation, he mentioned his work as the head coordinator of intercontinental ballistic missile (ICBM) tests between Vandenberg Air Force Base in California and Kwajalein Atoll in the Marshall Islands, a distance of almost 8,000 miles. Mr. Sudbury kindly agreed to a later telephone interview about his work, which we scheduled for May 22, 1998.

During that interview, Mr. Sudbury explained that the United States ran a *time* and *day* test over the new year date change in 1997 (it was not intended as a Year 2000 test). Uncharacteristic for U.S. testing schedules during the holidays, officials were looking to test clock rollover, 24:00 to 00:01 systems.

My notes continue with Mr. Sudbury's description of the issues with distance acceleration and time deltas, J. D. (Julian Date) 365 to J. D. 1 change, and use of both Greenwich Mean Time and Universal Time, mentioning that day and time are always a concern by virtue of the way we keep time. He related that the U.S. government collected more data than any other entity on Earth, stating that an encyclopedia volume of data is collected on those thirty-minute Vandenburg to Kwajalein tests.

Because he noted the sensitivity of his remarks during our conversation and my parallel reaction twenty-two years later when reading my notes, we must keep the details scant here.

Memorable during our conversation, Mr. Sudbury told me that in all his years of work, he was never able to determine from either a job interview or from a working relationship who would react best under pressure, in a crisis.[3] His words reminded me of the "Greatest Generation" and the legion of stories of the least likely soldier to perform the heroic act. It is sobering, isn't it? I had never heard anyone else mention that.

ISSUES AND PROBLEMS: THE WORK OF CAPERS JONES

Like any global problem needing a solution, there were knowledgeable, interesting, and sometimes amusing solutions offered for the millennium's computer compliance.

Capers Jones became known as an expert in Year 2000 issues and analysis, as chief scientist of Artemis Management Systems, Inc., and its subsidiary, Software Productivity Research, Inc., of Burlington, Massachusetts. He has authored and published a number of detailed abstracts of his research and projections of best-case, expected-case, and worst-case situations, costs, and consequences. My files contain fifteen of those abstracts, which were important factors as the country and globe prepared for the new millennium. Jones indicated that his clients included 200 of the Fortune 500 companies and 400 other organizations and government agencies. Because Jones was so often cited in articles and media on Year 2000 software issues, it is important to be reminded of what a credible expert was advising the country and the world.

In Jones's several February 1998 abstracts, he offered a chart of "Year 2000 Damage Probabilities Assuming Latent Date Problems," while noting a high margin of error. Topping the list of thirty-two problem sectors on this fairly early article (including manufacturing shut-downs, tax reporting, shipping, food and water shortages, etc.) was a 70 percent probability of bad credit reporting. Interestingly, next on the list was 60 percent probability of cancellation of Year 2000 liability insurance. Loss of local electrical power of more than one day was third at 55 percent. Last on the list was death or injuries, at 1 percent.[4] Later abstracts offered more detailed estimates within specific discipline sectors, with Jones's Version 3 (February 1999) abstract holding with the same probabilities as noted from his 1998 abstract.

Jones's original February 6, 1999, "Abstract" examined the legal issues arising from Year 2000, estimating that potentially more than 50 percent of U.S. directors and officers might be at risk, making Y2K the "most significant legal problem in history in terms of the number of lawsuits that might be filed." It was no secret that the plaintiffs' bar was gearing up for this, as discussed in chapter 5, with $1 trillion in litigation costs being widely published. Corporate America was justifiably nervous. In this abstract, Jones outlined the causes, percentages, and consequences of noncompliance within industry sectors most affected by the problem:

- Manufacturing: embedded software within physical devices
- Process Control: most manufacturing sectors including pharmaceutical and chemical

- Payroll: paycheck and tax withholding calculations, fines, and penalties
- Financial and Accounting: cash flow disruption and billing problems

The consequences of Year 2000 errors leading to litigation or even criminal charges were described:

- Americans and U.S. businesses affected by loss of electrical power and telephone disruption
- Death or injury from faulty products such as medical instruments, aircraft, weapons systems, and many manufacturing products
- Banking and security applications worldwide

Examining "an interesting matrix of four cells" of executive response followed, with an outline of corporate management rationale for both addressing and not addressing attention to Year 2000 issues. Assessment of compliance costs exceeding damage costs, as well as shareholder litigation from perceived excessive remediation expenses, were discussed, including charts of most remediation and damage expenses exceeding $100 million to $1 trillion. Of course, a short "incidental" mention of directors and officers liability insurance policies containing Year 2000 exclusions was mentioned as a possibility, adding considerable risk for executives.[5]

Capers Jones's volume of work on a wide range of Year 2000 issues became an important benchmark and planning guide for Y2K managers and corporate executives. His "Executive Risks" abstract of February 18, 1998 examined in detail the categories of liabilities corporate officers would be facing for Year 2000 remediation failures. From D&O liability to professional malpractice for corporate risk managers, attorneys, controllers, and auditors, Jones examined causes and consequences. He also warned, "[S]ome insurance companies have stated that Year 2000 damages would not be covered under existing director and officer liability policies."

Jones cited his 1997 interviews with 100 company executives and concluded that 35 percent of U.S. executives were "Year 2000 active" with 65 percent being "Year 2000 passive." Aggregating the combined costs of Year 2000 repairs, damages, and litigation could "top three trillion dollars on a global basis," the highest estimate of the time.[6] Disturbing conclusions for corporate officers and directors in the midst of analyzing and remediating both software and liability risks.

In his abstract dated September 1998, Jones analyzed the software release error rate over the fifty-year history of computer software existence, identifying the metric "defect removal efficiency." He measured an 85 percent rate between 1985 and his 1998 abstract date, applying that percentage to consequences of Year 2000 removal success and failure.

I should mention here that Jones noted that IBM projects in the 1970s had a near-perfect 100 percent defect removal efficiency but were not part of his study due to this being earlier than the formation of his enterprise. My own experience confirms his observation, as my career in computers started in 1970. I observed as the decades progressed that while IBM tendered seamless operational performance to its customers, Digital Equipment Corporation later was successful selling DIY computers designed specifically for adaptation by end users. Then came Microsoft, releasing bug-filled software, expecting early users to identify bugs for ongoing corrective updates.

Returning to Jones's September 1998 abstract; it charted *best-case, expected-case*, and *worst-case* scenarios for Year 2000 date failures, estimating an expected 12,000,000 software applications. Jones's late 1998 assessment estimated a little less than $500 million in *expected-case* costs, with over $1.3 billion *worst-case* costs, acknowledging a very large margin of error for his speculations. Worst-case Y2K issues were identified as possible lack of medical services, transportation, and food distribution, resulting in hoarding, runs on banks, riots, and civil unrest, with a ten-year economic effect from lawsuits, bankruptcies, and business failures.[7]

Jones's September 17, 1998, "Abstract: The Aftermath of the Year 2000 Software Problem" examined software remediation; best case 95 percent, expected case 85 percent, and worst case 75 percent, including the consequences of each scenario. Lawsuits, damage costs, U.S. per capita costs, power, transportation, communications failures, bankruptcies, unemployment, and associated stock market reactions were precisely estimated.[8]

"Version 3" of Jones's February 6, 1999 "Abstract" also included his assessment that almost one-third of all Y2K issues considered fixed for software problems found and repaired "early" were having date problems. Jones also made particular note that there were exposure issues, where management refused to provide data to him following his discussions with Year 2000 repair and risk executives, economists, and others. Specific also to Version 3 was a statement by the CIO of the U.S. Internal Revenue Service that the IRS would not be ready for Year 2000.[9]

In his forty-five-page, March 19, 1999, Version 2.1 "Abstract: Year 2000 Metrics," Jones enumerated a comprehensive list of documentation that should be kept for the specific purpose of litigation defense. I will not itemize the exhaustive specifics of this important work, but will note some interesting observations. Jones noted that with all the mission-critical necessity of Y2K compliance, software managers balked at monthly status reports, indicating they represented a lack of trust, answered by Jones to be no different than the detailed records kept by physicians and attorneys. A fascinating example of twelve different dating formats indicates the lack of standards of those past eras; U.S. formats, Julian format, ISO formats, "Dates in C applications if 100 appears for 2000 tm year(100)," encrypted dates used in military and intelligence applications, and other, more arcane date coding systems. This specific "Abstract" offers details under the headings of "Year 2000 Metrics for Criticality Analysis" (mission-critical applications), "Metrics That May Occur in Year 2000 Litigation," "Quantifying the Year 2000 Universe," "Metrics for Test Certification Criteria," "Metrics Analysis for Year 2000 Test Tools," and "Metrics Analysis for Contingency Planning" with references and suggested reading sources. Noting that by then, forty Year 2000 lawsuits had already been filed, Jones offered a twelve-month planning calendar beginning January 1, 2000, for damage control, litigation, and recovery.[10]

Jones noted the importance of a detailed remediation strategy for both insurability and litigation defense, likening it to "keyman" life insurance physical examination.

Jones's April 1999, Version 5 of "Year 2000 Contingency Planning for Municipal Governments" provided a detailed outline plan for municipalities' millennium remediation. Contingency planning, a monthly operational calendar 1998–1999, detailed 2000 emergency response measures, and his usual recommended reading list were included to prepare both large city and local governments for the millennium.[11] Excerpts from this abstract are referenced in chapter 6, "Legislation and Government on Year 2000 Issues."

In the May 1999 updated "Abstract," Jones suggested game theory (the science of logical decision making in humans, animals, and computers) as applied to the Year 2000 issue; either the remediation costs would be far less than the ultimate unremediated damage, or costs greatly exceeded what turned out to be minor inconveniences. Essentially, either fairly minor or very severe, as infrastructure noncompliance would affect human safety. Jones estimated a likely range of 36,000,000 software applications in the United States, again with *best case*, *expected case*, and *worst case* being 25 percent, 50 percent, and 75 percent Year 2000 "hits." Repair estimates of *best case*, *expected case*, and *worst case* were 95 percent, 85 percent, and 75 percent, respectively.

The abstract went on to identify eight categories of software applications, from management information systems (MIS) to military, with probabilities of failure rates being highest for end-user (35 percent) to military applications in the single digits.

Jones rated Europe for readiness, citing Germany and the United Kingdom as having the highest number of unrepaired systems due to the European Union's focus on implementation of the Euro currency. Interestingly, he stated that the EU in 1999 deployed slightly more software than the United States. A brief look at the Pacific Rim and South America indicated noncompliance projections having a grim outlook for their and the world's economies. Concluding projections of noncompliance consequences, best to worst cases, included no impact (best case) to economic depression, political crises, martial law, infrastructure collapses, and possible famine (worst case). With this important mathematical projection, Jones identified how the globe might be impacted by the millennium date rollover, January 1, 2000.[12]

Capers Jones's work was of great value to both those laboring for Year 2000 compliance and those managing the process.

A critical concern was hidden microchips, known as embedded microcomponents, which were used in a wide variety of consumer and commercial goods, and rarely used in computers. Some 4.6 billion micro*components* were manufactured in 1997 as compared to 100 million micro*processors*, which were used to perform computer computations. Embedded microchips were being used in everything from automobiles to microwave ovens, washing machines, factory controllers, to medical devices. Texas Instruments (where I once worked in IT) was a leading maker of these chips, known as digital signal processors, used in cell phones and consumer electronic devices. An estimate of the average number of chips in the American home in 1999 was 63, with projections to 280 in five years by Tom Starnes, analyst at Dataquest.[13]

The Food and Drug Administration reported that medical device issues with embedded chips caused considerable concern. A Hewlett-Packard defibrillator was identified as having a "set clock" message instead of the correct in-use time, and Invivo Research, Inc.'s patient monitors were also having a clock issue. The FDA was working with seven manufacturers of fifteen medical devices on Year 2000 remediation issues.[14]

OTHER REPORTED PROBLEMS

To include just a couple more relatively significant issues leading up to the millennium in this chapter: First, in 1993, a Year 2000 test was conducted by the North American Air Defense Command (NORAD) in Cheyenne Mountain, Colorado. Everything froze, including all early-warning satellites, radar systems, and systems detecting incoming missiles or bombs. Again, this was a simulation test, but it triggered an all-hands-on-deck Y2K remediation effort.[15]

The next was a *New York Times* article reporting on an issue named the "Crouch–Echlin Effect," named for Jace Crouch, a history professor in Michigan, and Canadian programmer Michael Echlin. Based on a design flaw in the basic input/output system (BIOS) of the battery-driven clock that keeps track of time when computers are turned off, clocks might possibly jump either backwards or forwards on January 1, 2000, called "time dilation." The *Times* article noted that extensive tests by both Intel and Compaq saw no evidence of the Crouch–Echlin Effect, even while Mr. Echlin was not conceding their theory. Disputes over whether or not the Crouch–Echlin Effect even existed went on for over a year, with critics arguing that the theory was created by two men selling solutions.[16]

Mark Anderson, in a September 20, 1999, *Boston Globe* article, quoted Thomas P. M. Barnett, senior strategic researcher for the Naval War College's Year 2000 International Security Dimension Project Report, with his experts as envisioning a "Global Rule Set."

> That is, if Y2K hits hard enough in enough corners of the globe, the report suggests that it could force global scrutiny on the new economy in a way similar to what happened at the end of World War II, which ended the isolationism of the 1930's and begat the United Nations, the International Monetary Fund, and the World Bank—the modern foundations of global politics and economics today.[17]

While Barnett went on to say that the United States is not enamored with the idea of a new "Global Rule Set," it would be necessitated by a need for change. "In short, its [*sic*] not what Y2K destroys that will be important but what it illuminates."[18]

These examples, among other real or imagined concerns, as well as Capers Jones's exhaustive metrices, offer some scope to the myriad of information deluging both management and labor as technology resources waded through to the successful conclusion of Y2K compliance.

PUBLIC RESPONSE

Solutions to the Year 2000 millennium date change issue covered a wide range of interests: computer software technology efforts, legal and management issues, litigation, as well as insurance coverage questions. Chapter 2 covered the technology software side, with chapter 3 covering corporate. This section includes citizen response as well as how Y2K litigation should be avoided with the use of alternative dispute resolution vehicles.

Humanity's panicked response to Year 2000 fears began to be a concern in the year running up to January 1, 2000. A *New York Times* article by frequent Y2K reporter Barnaby Feder in February 1999 suggested those fears might lead to "bank runs, hoarding of food and gasoline, fires caused by misuse of newly acquired wood stoves and generators, and a rise in gun violence stemming from the surge in firearm sales to those fearing civil unrest." Hoarding money was suggested to lead to loss by theft, scammers, and binge shopping.[19]

Citing nearly half of local governments and small businesses as not having started on Year 2000 remediation efforts, John Koskinen, the nation's Y2K czar, noted: "For some people, a certain amount of panic would help." I do want to note that Feder reported in the same article that organizers in Spokane, Washington, called for a "practice day," asking residents to do without the basics of electricity and running water.[20] I wish those of us in Concord, Massachusetts had been aware of Spokane's test at the time; it would have been great to hold a "practice day" here as well.

A lengthy September 1999 *Boston Globe* article described the "first staged mock Y2K chaos" five-hour test a year earlier in Lubbock, Texas. Emergency management officials ultimately testified before Congress on the importance of emergency preparedness. The same article mentioned a "small preparedness group" located in an unincorporated area surrounded by concertina wire (curled barbed or razor wire) near Spokane, Washington, noting that any member of the group speaking of it would be thrown out.[21] More practical readiness in the same *Globe* issue related stories from New England; an East Machias, Maine resident featured in his five-holer outhouse, as well as describing preparations for water and electricity. Stories of others in the northeast noted buying lanterns, guns, extra food, and stockpiled water. A discouraging item was included about a Connecticut church council who tried to persuade members to "buddy-up" with fellow parishioners in the event of Y2K infrastructure failures, with little success.[22]

"Pogo's Revenge Remembered" needs to be recalled for those of us who remember "We have met the enemy and he is us," quoted so frequently over time. And for Year 2000, too. The *Times* article by Feder suggested that the unusual combination of "human foibles like sloth, greed and unfounded optimism that someone else—or new intelligent machines—would take care of the problem allowed it to grow into an expensive potentially deadly global challenge." Feder suggested some Americans saw the issue as a divine punishment, with survivalists fleeing for rural fortresses, asking "Will citizens be neighborly, rather than selfish?" Ten percent of Americans planned to withdraw their money from banks, with polls indicating a growing number of people planning food stockpiling and other backup precautions.[23]

LITIGATION ALTERNATIVE SOLUTIONS

As the prospect of massive, crippling litigation loomed over corporate America, efforts to find alternatives began to be seriously presented. The prospect of national property/casualty insurance market bankruptcy and collapse suggested alternatives to litigation were not only needed, but required. I was among several voices calling for alternative dispute resolution and mediation as solutions to costly litigation. While legislated protections for best-effort Y2K compliance measures helped, litigation refuting "best efforts" would be inevitable. Corporate back-offices laboring in America, law firms eager for new litigation opportunities, and insurance executives bracing for massive hits needed to find cooperative common ground for January 1, 2000. Remember that they, and every sector of the country, were also laboring in their own offices to be Year 2000 compliant.

The *Insurance Times* published a guest editorial of mine in their September 15, 1998, issue.[24]

A Modest Year 2000 Proposal:
A Millennium Mediation Council
by Nancy P. James

Having just returned from London, where the EU insurance community asked for a strong U.S. lead on Year 2000 coverage concerns, I am offering a modest proposal which breaks all our traditional business paradigms. Just a year ago, on August 1, 1997, I addressed the American Bar Association Annual Meeting on the Year 2000 subject, concluding my early predictions on Year 2000 influences by asking the bar to join, as they have never before done, with their insurance and accounting professional colleagues in solving the Year 2000 problem for our clients.

Today it is a categorical imperative! Current estimates of this $600 billion global problem will exhaust twice over the U.S. estimated $333 billion domestic insurance reserves without even touching the $1–$1.5 trillion the bar expects in litigation costs and fees.

The EU community, without equivocation, concurs that U.S. courts will likely deny our Year 2000 exclusions. Exclusions denied assume coverage on the basic, unendorsed policy. Letters of clarification from carriers go a long way toward client understanding of coverage, but ultimately, no affordable independent commercial Year 2000 coverage exists in the states today. And, coverage areas have yet to be tested in the courts, which, in my judgment, will rival patent litigation costs of $500,000 pre-trial and $500,000 at trial.

We still have an opportunity to solve this situation before Year 2000 arrives, and allocate every precious available resource toward the solution and away from adversarial dispute expenses.

A national body, *The Millennium Mediation Council*, including insurance, accounting, and law professionals must settle coverage questions and limits ahead of time; mediation before the fact for this unfortuitous event. Claims limits within categories need to be resolved, as does uncovered areas of Year 2000 loss, attorney's fees and stated, common defended areas (fire, bodily injury). These areas can be quantified; we can advise our clients precisely; and we can be a valuable part of a global solution to Year 2000. Thus, we will be able to direct U.S. resources toward technical Year 2000 solutions and fail safes. We can lead the EU and global insurance communities in a rational strategy.

And, we must urge our industry and government officials to support us in our efforts.

After the publication of my proposal, I received a letter from Sandra A. Sellers, president of Technology Mediation Services, LLC (McLean, VA), who had recently published a similar suggestion in *Legal Times* the previous month.[25] While the position in my article was clearly pointed toward difficulties and costs involved with insurance disputes, Sellers' article was directed toward saving of litigation expenses and the risks associated with uncertain outcome of litigation. Sellers raised an interesting point that mediation, unlike litigation, safeguards confidentiality and reputation, both essential to ongoing company viability.[26]

Sellers pointed to the first Y2K lawsuit, *Produce Palace International v. Tec-America Corp.* in June 1997, which involved the failure of cash registers to recognize credit cards with expiration dates of 01/01/2000 forward, resulting in a measurable loss of income for the retailer, which I noted in chapter 5. Sellers described the case as being referred to "mediation," a mini-arbitration, in February 1998, ultimately resulting in a panel of three practitioners recommending a $260,000 settlement, which the plaintiffs refused. The parties then agreed to "facilitation," a procedure closer to mediation.[27] The ultimate outcome of this first Y2K legal case, closely watched by interested parties from many industries, was met with varying satisfaction.

Insurers were especially interested in a nonlitigation solution, arbitration or mediation, not only for economic motives, but for preservation of relationships with clients and business partners. Janet Bachman, a vice president of the American Insurance Association, was quoted in *Best's Review*: "'Litigation is by far the slowest and most expensive option.' And, she pointed out, '[I]f a dispute is settled through negotiation or mediation, there's no possibility of unfavorable case law precedents being set.'"[28]

Lisa Romeo of the American Arbitration Association offered comments mid-1999 regarding the range of litigation expected from Y2K issues, from product failure, to supply chain disruption, corporate officer liability, and insurance coverage disputes. Noting already overcrowded courts, Romeo added to a growing call for arbitration and mediation, offering flexible and attractive alternatives to litigation. Both arbitration and mediation are known as Alternative Dispute Resolution (ADR), and involve an experienced neutral party to sort through the technical and legal issues. She noted the distinction between court procedures, which primarily determine liability and damages, to ADR, which is tasked with determining the source of the conflict and assisting in finding jointly agreeable resolutions. ADR is consensual, confidential, faster than court proceedings, and cost effective, with the ability to tailor the proceedings to fit both parties' needs. Romeo supported nonbinding mediation between parties wishing to continue business relationships as the ideal tool, with arbitration being the final binding resolution if parties were still in disagreement.[29]

Other powerful voices began advocating for cooperation instead of litigation, often with differing approaches. Danny Ertel and Jeff Weiss in a "Public Forum" article published in the *Boston Globe* recommended the Y2K issue be approached with a cooperative spirit of partnership. Their editorial concluded with:

> The Y2K problem is only partly about technology. It's also about how to do business in an interconnected world that is highly dependent on information and company-to-company relationships. Consider the choices: Spend millions of dollars in litigation that will at best provide inadequate monetary compensation, years after the fact, to businesses already badly damaged. Or approach the problem in a way that acknowledges those relationships to solve the Y2K problem expeditiously and collaboratively.[30]

The U.S. Chamber of Commerce in late 1998 recommended a special court be established to handle Y2K disputes. "'We are concerned that Y2K-related litigation could virtually shut down a part of our economy,' said Larry Kraus, president of the U.S. Chamber Institute for Legal Reform in Washington. 'This Y2K court could sunset itself in five years after the turn of the century.'" However, the National Association of Manufacturers (NAM) was not ready to endorse a special court, advocating for congressional processes for Y2K issue resolution. The Information Technology Association of America considered the use of a special court problematic since technology issues unrelated to Y2K might be unfairly settled there.[31]

A *New York Times* letter to the editor in late December 1998 from the CPR Institute for Dispute Resolution president, James Henry, took issue with a *Times* article reporting on law firms "gearing up for a litigation wave of huge proportion." Henry argued that attorneys were working hard to avoid litigation, with corporate signatories promising negotiation and mediation.

Sophisticated business lawyers, he stated, recognized the importance of relationships and understood the need for cooperatively working together on Y2K issues.[32]

Other organizations and trade groups were supporting alternatives to litigation when the Gartner Group published in January 1999 that 38 Y2K lawsuits had been filed with 711 more in the pre-filing stage. Technology and manufacturing companies, anxious to avoid court, represented by the Information Technology Association of America and the National Association of Manufacturers, supported congressional legislation to limit liability. The American Trial Lawyers Association, asserting such legislation was "simply tort reform in sheep's clothing," vowed to defeat congressional proposals.[33] Lines in the sand were deepening; hopes for cooperative effort dimming.

Interestingly, those Y2K watchers (who did not have a pony in the race) suggested that the best way to allay citizens' fears would be with constant information to the public on Year 2000 remediation and compliance. That advice ran contrary to the concerns of municipal, public, and private entities' fear of litigation as a consequence of unintended misinformation during the process.

Opinions from corporate and public interests often run contrary to one another, then and still. A common goal of success remains on the bottom line of each. Cooperation and common sense occasionally prevail. It did for the great millennium Y2K challenge!

8

INSURANCE AND Y2K BEYOND THE UNITED STATES

I was brought to London on July 22 to 23, 1998, by an invitation from the International Quality and Productivity Centre, Ltd. (IQPC), to speak at their Fourteenth Annual Millennium Management Conference titled "Managing Internal and External Exposure to Year 2000 in the Insurance Industry." I was the only speaker from the United States. My topic, "Wild Card Projections," was scheduled for the very last session of the two-day conference; the promotional material text of my remarks appears below. Quite frankly I had every expectation that, as a speaker on the last day of the conference, my entire audience would be comprised of my husband, the conference chair, and perhaps one or two of the other afternoon speakers. Yet when we arrived, the room was filling with people. I began to think perhaps I might have asked more specifically why I had been invited. To say the conversation was lively is to understate it.

At that time, I was close to a veteran reporter, Penny Williams, from the Boston-based *Insurance Times*. I sent her this full report of my experience, which appears below. My "Report from London" is the best reconstruction I have of my experiences.

4.30 WILD CARD PROJECTIONS

N P James Insurance Agency in Concord, Massachusetts, USA, specialises in risk analysis and insurance for technology-based clients. The agency has years of experience in understanding products liability, errors & omissions liability, directors & officers coverage and other key issues unique to technology risks. From this background, Nancy P James will cover the following issues:

- US courts intervene to deny Year 2000 policy exclusions
- Governments intervene to help businesses cope with Year 2000 shutdown losses and inadvertently trigger unintended insurance coverage
- Professionals may face our own uncovered liability exposures with Year 2000 exclusions. How anxious will we be to work with clients on their Year 2000 problems?
- Will very large law, accounting, and insurance firms, who are able to leverage professional E&O coverage for themselves for Year 2000, put smaller firms out of business?
- Will the soft market turn hard?
- Will our uncovered clients hold a grudge?

Nancy P James, *President*
NANCY P JAMES INSURANCE, USA

Courtesy of the author.

N. P. JAMES INSURANCE AGENCY
33 BEDFORD STREET
CONCORD, MA 01742
TELEPHONE (978) 369-2771 FAX (978)369-2778
NPJAMES@COMPUSERVE.COM WWW.NPJAMES.COM

REPORT FROM LONDON

"YEAR 2000 IN THE INSURANCE INDUSTRY"

July 22, 23, 1998 Sponsor: International Quality and Productivity Centre, London

Penny Williams—some of this is extremely controversial: we probably do not want to run it as is; DO NOT SUBMIT, PLEASE, WITHOUT SPEAKING WITH ME. But this is what they said [Zurich Re would not allow reporters, so excluded].

The management of both internal and external exposures to Year 2000 were the topic of discussion among an international gathering of insurance executives in London July 22 and 23, 1998. This report will cover the matter of Year 2000 coverage considerations, where speakers have not expressly forbidden disclosure. Attendees and scheduled speakers from 38 insurance, law and business professions, included representatives from AIG Europe, A.M. Best Europe, Deutsche Bank A.G., Green Flag Ltd./UK Insurance Ltd., J&H Marsh & McLennan UK, Prudential

Insurance, Reinsurance Australia, Royal & Sun Alliance, Tillinghast-Towers Perrin, Zurich Insurance Ltd., and Zurich Re (London) Ltd. Nancy James, N. P. James Insurance Agency, Concord, MA, USA, who addressed the IQPC conference on the current U.S. insurance industry response, was the only American attending.

The opening remarks by Peter Worrall, Editor of the UK *Insurance Specialist*, included the statement that "if America goes down, we go down." James was asked on several occasions why the United States was not taking a strong lead on coverage matters, which the EU community thought critical to preventing going down. Chris Waterman, A.M. Best Europe, looking at the company rating process stated that no companies had to date been downgraded for Year 2000 noncompliance or risk exposure. Best's is looking at this time for a pro-active approach to Year 2000, without offering direct recommendations. Waterman also stated that Lloyd's had already received Professional Errors & Omissions claims from systems providers, and predicted that insurers would be using Year 2000 exclusions at renewal.

The use of exclusions came under vigorous discussion during the open forum portion of the conference, with one reinsurer standing firmly opposed. James argued that the magnitude of the problem precluded any solution but to discourage the bringing of Year 2000 remediation costs and damages to the insurance industry. One attendee representing the public sector stated that Year 2000 problems would be suppressed during the American presidential campaign of Vice President Gore for political advantage purposes.

Paul Smith, Underwriting Director from Green Flag Ltd./UK Insurance Ltd., and transportation and travel insurer discussed the implications and public expectations of travel and transport claims coverage. Smith stated that the public currently had a "reasonable expectation" that losses would be covered, including consequential damages, and made some chilling statements that 50% of all computers sold world-wide in 1997 were not Year 2000 compliant, and that moral hazard losses were already being seen as consumers were destroying non-compliant insured systems to replace with Year 2000 compliant ones.

Lisa Hansford-Smith, J&H Marsh & McLennan Y2K Team (UK), reviewed the legal, accounting, regulatory, and litigation climate surrounding Year 2000, stating that Year 2000 was not only difficult to underwrite, but unusual to the problem was the fact that insurers have no time to build premium for claims payment. Warning that hackers would be ready to loot assets from chaotic systems situations, and viruses causing rampant destruction, Smith described in detail Ascent Logic's services to J&H clients in consultation and commercial remediation services to insurers, and secured website to subscribers. When asked if any J&H Year 2000 policies had been sold in the United States, Smith replied that she knew of none, but that Ascent's services have generated considerable interest and inquiry. Smith noted that in the United Kingdom third party audits were not as common as in the United States, and for that reason Ascent's services were of distinct value.

Adam Taylor, law partner and intellectual property and information technology specialist, Withers, covered detailed contractual obligations and exposures of technology suppliers, contractors, and consultants, with primary focus on a less burdensome climate of the UK and EU communities.

Michael Graham, Partner, Barlow Lyde & Gilbert, legal counsel to insurers and reinsurers worldwide, addressed the conference on mitigation of exposures to Year 2000 claims. Graham focused on four key strategic issues: internal systems compliance; trading partner system compliance; development of a coherent underwriting strategy in a soft market; development of a consistent reserving methodology. Defenses against coverage included willful misconduct for compliance failures, fortuity issues, and non-disclosure of material facts to insurers. Possible disputes between 1999 and 2000 insurers were noted, with exposure mitigation including questionnaires, insured's written Year 2000 compliance warranties, material development disclosure requirements, Year 2000 policy exclusion attachments, and possible policy cancellation provisions. Offering specific legal advice to insurers and reinsurers regarding policy clauses and wording, Graham concluded with his opinion that "Y2K Prudent Underwriter" will survive the "Market Share" or "Must-Do But Cannot Decide" positions prevailing in today's market.

Adrian Smith, Risk & Insurance Manger, Excel Logistics, followed with "Customer's View," reviewing in detail Excel's remediation efforts for Year 2000, both for its own plant as well as supplier and customer issues. Smith remarked with respect to insurance coverage matters that Year 2000 was not an insurance issue. Time and money were better spent on fixing, testing and replacing non-compliant systems than in purchasing insurance coverage, customarily requiring expensive external audits. Year 2000 could be a blessing in disguise, Smith stated, offering businesses the opportunity to replace and refresh technology components and systems with improved capability, improve the quality of external relationships, auditing, and procedures. Negative impacts of cost and operations interruptions are outweighed by the opportunities.

Nancy James, N. P. James Insurance Agency, Concord, MA, the only American speaking or attending the conference, provided the U.S. perspective. The EU community expressed uniformly its hope that the USA would provide a strong lead in Year 2000 coverage matters, and its disappointment that, in James's words, only a "roaring silence" could be heard to date. The text of James's remarks are covered in this issue under the title "____________."

by: Nancy P. James

The aforementioned review of the program discussions displayed an unpleasant example of sausage being made by financial professionals. Estimates of the magnitude of the Y2K millennium compliance, remediation, and consequential litigation costs would surely destroy the casualty/property insurance market in the United States and the European Union should those costs be shouldered primarily by the insurance sector. The breadth of opinion, mixed with self-interest, was more openly displayed in that London conference room than I have ever experienced before or after. Reporting this as I had described would only add to the anxiety in the U.S. market. It was then, and remains, my opinion that little good could come from such an exposé.

To highlight how politics plays a significant role in humanity's best efforts, a noteworthy article appeared in the British publication *Business* in mid-1998 during Parliament's transition from Conservative to Labour Party, revealing that Robin Guernier, the UK's chair of Taskforce 2000, was relieved of his duties. Labour was turning its attention from Y2K awareness to solutions with a new organization called Action 2000. Guernier had fallen out of favor by criticizing government's Y2K efforts and, as a consequence, his original Taskforce 2000 lost government funding support. Thus, even with Prime Minister and EU president Tony

Blair prioritizing Year 2000, Guernier concluded that both the private and public sector's efforts were too little, too late.[1]

LONDON INSURERS SEEMED TO WORRY MORE THAN U.S. INSURERS

"Insurer Liability Warning on Millennium," a December 11, 1996, article in the UK publication *Insurance Day*, was a relatively early warning in the UK of dire predictions of huge insurance claims for Year 2000 failures. Not only did insurers need to bring their own systems into date compliance, but faced claims from any number of fronts, including computer and software vendors, corporate directors and officers, accountants, and others.[2]

In early 1998, *Insurance Day* published another article about UK and EU courts, suggesting that courts could challenge Year 2000 policy exclusions. If so, insurers would be open to massive claims.

"Lawyers say courts could rule that Year 2000 exclusion clauses are in breach of the provisions of the Unfair Contract Terms Act (UCTA) 1977, which deems that any exclusion clauses have to be reasonable, in the opinion of the court."[3] Stephen Digby, a solicitor with London law firm Withers, noted considerable concern over the UK courts' response to Y2K policy exclusions. Digby continued: "The risk for insurers is that if Year 2000 exclusions are considered to be unreasonable by a court then it has the power to delete them. Therefore, any limitation or exclusion of liability is ineffective. . . . In any such a case . . . the court may decide to impose a limit on an insurers [*sic*] liability, or simply say that is [*sic*] liability was unlimited."[4]

In a London *Sunday Times* article in late July 1998, Robert Winnett offered harsh criticism and commentary on insurers' Year 2000 exclusions, describing a host of consumer appliances that might suffer millennium change failures. The first item on his millennium checklist advises consumers to contact their insurance company or broker to find out what Y2K issues were covered (then still a huge unknown, which had been widely publicized). Of note was his reference to the UK's 1979 Sale of Goods Act, which provides full refund good for up to six years following purchase if Y2K failures occur. Winnett also offered a rather astonishing quote from a Microsoft spokesman: "We have little software fixes we can send to people, or that can be downloaded from the Internet by users to sort out any problems, so there is no need for them to worry unnecessarily about their personal computers crashing."[5] The article clearly pointed out a noticeable disparity in the UK between the public's and the government's approach to Y2K issues.

By mid-1998, UK insurers led those in the United States in adding Y2K exclusions to their policies, according to Rosalind Jones, a solicitor with the London firm Elborne Mitchell. The position of U.S. insurers at that time was that policies did not cover Year 2000 problems in any case (see chapter 4, "Year 2000 Is Not Covered"). Aon was cited in a mid-June 1998 article stating that the firm had written underwriters in the London market to inform them that Aon clients would not accept such exclusions.[6] Aon was stating to the insurance world, gratuitously in my opinion, that it would not be permitting Y2K exclusions on its policies.

Aon made the same statement at the IQPC London Year 2000 conference where I spoke on July 23, 1998. Upon returning to the States, I asked my underwriters if Y2K exclusions were going to be used, and would they be used uniformly or favor large brokerage houses. Their answers were "absolutely uniformly." Among the many quarrels and disputes heating up around Y2K, I considered such posturing on Aon's part as not more than a grab for market share. In the world of insurance, all terms and exclusions may be negotiated, but always with risk assessment and appropriate pricing; an underwriter's decision, but ultimately the insured's choice to accept.

London consulting group BSC published an attractive information sheet reviewing Year 2000 issues, including detailed embedded systems charts and brief air traffic safety and insurance comments. BSC, like U.S. commentators on insurance costs, estimated a 10 percent premium to limits rate, as well as the necessity of expensive compliance audits. Derek Brighton of the UK's Association of Insurance and Risk Managers offered the position that insurers would not respond to Y2K claims, stating, "Act as if you are uninsured. The insurance industry *has* to take this view to protect itself. After all . . . they did not plan for the Year 2000 and they should not be expected to carry the can for it either."[7]

It was clear at the time that British insurers were very much more worried about the impact of Y2K claims and the UK courts than were Americans. However, most of the articles in my collection are written by attorneys whose own liabilities for Y2K were never addressed. By contrast, a May 1998 lead article in *The Standard* expressed brokers' frustration that insurance companies had not to date indicated how Year 2000 issues would be treated.[8]

GLOBAL WORRIES VARY

With the storm of articles written about both U.S. and global Y2K readiness, opinions varied widely, even dizzyingly. A January 1999 Reuters article reported that the head of Swiss Re America, Chief Executive Heidi Hunter, was concerned that "the United States was not prepared to cope with the Year 2000 computer bug, warning of potential harm to the nation's economy and civil order." Hunter added that the U.S. federal government had "not done enough to make sure the nation's critical power, phone and banking systems function properly when the new year arrives."[9] In contrast, a *Best's Review* article reported that "outside the United States, the Y2K effect could be worse, since most studies reveal the United States and Great Britain to be the most advanced in the Y2K conversion process with Europe primarily focused on its currency conversion and Asia embroiled in financial crisis."[10] The U.S. Senate testified that the consulting group Bond reported that France and Germany, leaders in the euro conversion, considered the unified currency more important than Y2K remediation.[11]

A very prominent organization called the Global 2000 Coordinating Group, consisting of 250 banks, securities firms, and insurance institutions from forty-six countries, was formed to collaborate on global readiness. According to a January 2, 1999, *New York Times* article, the Group's meeting deliberated on the hazards of rating countries on Y2K preparedness, fearing for the consequences to lower-ranking countries. Such public rating would not only affect financial instability, but would subsequently result in the Group's lack of further access to countries' national infrastructure data. Meeting every six weeks, members ultimately decided not to publish country readiness ranking. The *Times* article concluded with a note that the Group's website (www.global2k.com) "bluntly warned that Russia will not solve its Year 2000 problems in time." A subsequent *Times* report of February 1, 1999, confirmed the Group's decision not to rank countries, but promised to share its findings with the United Nations and the Joint Year 2000 Council, an international coordinating committee.[12] The Gartner Group at the same time was rating eighty-seven countries on Y2K preparedness.

The World Bank, the OECD (Organisation for Economic Co-operation and Development, an international organization), numerous multilateral development banks, and international private sector organizations undertook global Y2K awareness as reported by James Bond, coordinator of Year 2000 operational initiatives at the World Bank.[13] Lending to developing countries for Y2K compliance was seen as a global necessity, since interdependence of basic infrastructure resources ultimately affects the entire international community.

Numerous international organizations joined to alert the world to the Year 2000 computer issue. A press release by the International Association of Insurance Supervisors stressed the "tremendous risk of disruption in the operations of financial institutions and in financial markets" in conjunction with the Basel Committee on Banking Supervision and International Organization of Securities Commissions (IOSCO, representing central banks from Europe, Canada, Japan, and the United States). The November 1997 press release urged members to access regulatory authorities' websites for information.[14]

Harvard researcher Juan Enriquez noted that "[T]he folks who manage the flow of world capital are worried about Y2K" and would be seeking safe harbors for their money. "This could have a devastating effect on developing countries. Massive institutional capital flight means simultaneous financial meltdown in many emerging markets." Enriquez predicted a boom in U.S. blue chip stocks and bonds as a consequence, while the developing world could bust. He also noted the flow of money from Mexico to the United States, and from Russia to Switzerland, urging Africa, Asia, and Latin America to take Y2K seriously.[15]

The United Nations convened a task force of over a hundred member nations in December 1998 for a Year 2000 remediation working session. With troubling reports of global ignorance and inattention, basic infrastructure issues emerged as critical problems, especially among U.S. trading partners. Predictions of lawsuits to U.S. compliant companies from noncompliant overseas third-party suppliers posed a real threat.[16]

The Joint Year 2000 Council of international banking, securities, and insurance sectors meeting in Geneva in mid-1998 issued a statement of global concerns regarding Y2K infrastructure readiness. In particular, they shared worries about utility failures, noting that consultations with utilities executives "were not reassuring." National coordination of Year 2000 readiness efforts was encouraged globally.[17]

Russia had over two dozen nuclear power plants operating since the mid-1980s, with growing global anxiety over Y2K operational issues following the 1986 Chernobyl reactor explosion as a consequence of inadequately trained personnel. A March 1999 *New York Times* article reported on a closed Senate briefing at the Capitol the previous day, where concerns were expressed over computer failures in Russia, which "could blind early-warning radar systems and lead to false alarms of nuclear attacks. Soviet-built nuclear reactors in Russia and Eastern Europe could abruptly shut down." In a rare mention of terrorist acts around the Y2K issue, the briefing noted a "low to medium" probability that terrorists would try to exploit the resultant chaos.[18] Also reporting the same day on the Senate briefing, a *Boston Globe* article cited Senator Robert Bennett, head of the Senate's Special Committee on the Year 2000 Problem, as saying the possibility of intercontinental warfare could not be ruled out, even while "There is a low probability of [an accidental] nuclear launch."[19]

A mid-1998 *Boston Globe* article reported that the United States was sharing early-warning satellite data with both Russia and China to prevent uncertainty and panic over the possibility of an unknown attack if Y2K darkened those countries. Bruce Blair, nuclear-weapons specialist at the Brookings Institution, interviewed Russian nuclear officers, reporting that Russia was relying far more on nuclear weapons following the collapse of its army. However, shortages in both labor and money had weakened Russia's nuclear weapons systems with vulnerabilities. The article also reported top Pentagon officials only learning of U.S. Year 2000 issues in 1995 from the Social Security Administration, then shifting $1.9 billion from weapons programs to Y2K.[20]

Ilana Gerard, trend forecaster and strategic planner, offered an overview of the world's Year 2000 readiness, noting both the United Nations and the World Bank's efforts to assist struggling countries. Multinational companies, precipitating the push for Y2K compliance often thwarted by local government bureaucracies, were assisted by the World Bank's $30 million in funding to 120 countries for foreign government and technology personnel seminars. Gerard predicted that severe Y2K disruptions to food and fuel, infrastructure, healthcare, education, and more could result in social instability, straining diplomatic relations, and returning some countries to a less technology-driven society. She concluded that the most successful course toward global stability is for international cooperation; the more prepared countries assisting the less prepared. In the face of wildly divergent and often loud opinions, Gerard bravely offered "A Forecast or Two":

- This is not going to be as bad as some people think.
- In the United States, it will take six months to a year to sort out most of the bugs. An occasional problem will surface from time to time after that, but most of the major problems will all be solved by June 2000.
- In other parts of the world, especially in emerging countries, the fix will take much more time, maybe as long as three or four years.
- The stock market will continue to be extremely volatile through the new year, and then by February will begin to stabilize and show steady and solid growth.[21]

The United Nations, in an Associated Press release in June 1999, reported on the second global conference on Y2K affecting 170 countries, including some closed-door session reports by attendees. Pakistan's UN Ambassador, Ahmad Kamal, chair of the UN Working Group, conveyed greater global optimism for readiness, stating that the initial focus on banking had moved to shipping, healthcare, and other sectors. Interestingly, the article noted "empty chairs at the conference including Samoa, Togo, Guyana, Haiti and the Vatican." Political pressures were also noted; for instance, Israel's UN Ambassador Dore Gold expressed distress that a Middle East regional Y2K meeting was not attended by Arab countries, as they refused to sit down with the Israeli delegation.[22] Despite the potential for catastrophic infrastructure issues, political rivalries did not disappear.

MISCELLANEOUS GLOBAL NOTES

A sample of quotes that were noticed by the U.S. media: "Iranians reportedly blamed Y2K on 'the Great Satan.'"[23] "Indonesian Cabinet ministers are on orders not to talk about it."[24] "Compliance remediation difficulties were said to be due primarily to the fact that 90% of Chinese software in use was pirated."[25] And predictably, "Israel's U.N. Ambassador Dore Gold lamented Tuesday that a regional Y2K meeting for the Middle East and North African countries never took place Monday because Arab countries wouldn't sit down with the Israeli delegation."[26]

FAILURE RATES PREDICTED FOR COUNTRIES BY THE GARTNER GROUP

In a lengthy late-1998 report from the Senate Special Committee on the Year 2000 Technology Problem, the Gartner Group's grid of failure rates by country was included. Lower failure rates unsurprisingly included many EU countries, Australia, Canada, and the United States. South American countries and Japan were at moderate risk, with African and Southeast Asian countries, and China and Russia, least prepared and at high failure risk.[27]

While there may be some curiosity about how accurate those estimates were, no one seemed to be paying much attention after January 1, 2000 arrived so wonderfully quietly. Certainly the thousands of documents in my files make no mention of it; but then, I had to get back to my work, too.

CONCLUSION

As I have stated, the great preponderance of my documents relate to U.S. efforts for Year 2000 remediation. These several notes of foreign readiness methodology serve primarily to explain the breadth of national thought, opinion, and attitude. Reading the above, one cannot help but wonder how it all worked so smoothly in spite of such substantial obstacles and diverse positions.

9

Y2K GAFFES, SCAMS, WIT, AND WISDOM

During the several years leading up to the new millennium, the usual pundits had a lot to say, from the frantic handwringers to the deniers. Documents saved from those years offer amusing reflections on how Americans mostly regarded the approaching Y2K as either Armageddon or a New Year's Day yawn.

One of the actions I took to relieve the pressure and blitz of Y2K-related unknowns was to create some Y2K Police cartoon postcards to send to my clients.

Courtesy of author.

THE PSYCHOLOGY OF Y2K

I have offered the comments of Ramsay Raymond in the introduction, referencing a culture held hostage by fear of litigation. Otherwise, little real perspective of the human psyche had been offered regarding the widely publicized preparations of apocalyptic survivalists. However, Bernard Kaplan, professor emeritus of psychology at Clark University in Worcester, Massachusetts, offered insights published in the *Boston Globe* five days before the new millennium. Kaplan explained that deep in the collective consciousness from earliest primitive societies was a latent fear of the end of the world every year as winter approached with shorter, colder days and dying vegetation. When significant events coincide, fear intensifies with a sense of threat to all values and aspirations.[1] Rituals are used to assure continuation of life and culture, in part explaining survival readiness, both normal and hyper. Of course it is within human nature to react with excessive anxiety, fear, and self-protection, as well as the opposite—to turn a blind eye to the perceived crisis.

Courtesy of the author.

UNUSUAL Y2K OCCURRENCES AND EVENTS

In an unusual mea culpa, Tim Graham, the editor in chief of *TechWeek*, apologized to his readers on June 29, 1998, for publishing "Doomsday Fears About Y2K." Graham noted that many of his readers had complained about the article's ominous prophesies, acknowledging that the article had not met his journalistic standards, and that "we should have presented a substantially more balanced story, with sober analysis from experts who are more sanguine about our chances of fixing Y2K programming glitches before it's too late." He concluded that "dire predictions tend to overlook the resiliency of the human spirit, the common denominator in every good disaster movie. What we don't know is, who will be the heroes in this real-life drama?"[2]

What is interesting in this apology is that in mid-1998, it was so insightful. My files are full of documents predicting a mid-winter Armageddon as infrastructures collapsed in North America, Europe, and around the globe.

Yet arguably there *were* some amusing occurrences. An October 1999 piece in the *Boston Globe* noted that the Maine Registry of Motor Vehicles were identifying 2000 model cars as "horseless carriages" since April. Maine used the identification for vintage vehicles produced before 1916. Then Governor Angus King was surprised by this when called by reporters, while banks and lienholders were requesting clean titles from car owners.[3]

Dating errors in early 2000 did not seem to hinder sent and received emails. In fact, I received a mid-January email dated 1/17/1000 (rather than 2000) from an attorney friend in Boston.

Unresolved: It was evident that Microsoft also briefly used 10-year windowing. My archived office documents contain files dated 1/6/1999, 2/23/1989, 12/1/1989, 12/2/1999, and 1/10/2000. Yet I know some of these dates are impossible—the 1999 dates were changed to 1989 in my archived file lists. This was only noticed when I embarked on writing this book. Unlike time and date sensitivity today necessitated by constant email flow, most

Courtesy of the author.

business users created and located documents via computer desktop file folders. Significant dates were located directly on the created documents, usually written to be posted by U.S. Mail. My inquiries to Microsoft have had no explanation.

OPINIONS OF MERIT . . . AND THOSE OF NO MERIT (*YOU BE THE JUDGE . . .*)

The following collection of quotes and anecdotes is offered without commentary. They are offered on a "you be the judge" basis as examples of the spectrum of opinion that will prevail during any large-scale crisis. Anticipating such diverse opinions from both experts and casual observers helps to manage any critical path forward. One might think of these as possible detours on the roadmap to a successful outcome.

As 1998 drew to a close, "millennium hysteria" became an often-cited buzz phrase. Simson Garfinkel, technology writer for the *Boston Globe*, looked back at his early 1998 reporting of his expectations that most would "probably be okay." Garfinkel's October 8, 1998, *Globe* column Plugged*In* reported a torrent of emails faulting him for not sounding the alarm, including one from Ed Yourdon, a well-known Y2K consultant, readiness critic, and author of *Time Bomb 2000*: "People's reactions to the potential crisis might cause as many problems as the crisis itself," pointing to the little-published issues of the psychological effects of millennium angst.[4] Certainly in 20/20 hindsight we know that society experiences reactions, such as the great toilet paper shortages during the COVID-19 pandemic crisis of 2020 and 2021. Garfinkel concludes his October column by quoting Year 2000 experts stating that both the "all is well" and the "doomsday" predictors were both likely wrong.[5]

An October 1998 *New York Times Magazine* offered a short piece citing Ed Yourdon and conservative millennium doomsday predictors Gary North, Pat Boone, and Tim Wilson, publisher of

Y2K News Magazine. The *Times* piece offered, "Depending on whom you ask, this techno-flummox is either a thorny but fixable problem or the First Horseman of the Apocalypse." Survivalists' readiness for the Y2K frontier was featured.[6]

Wired magazine ran a feature article on Scott Olmsted, a Y2K software engineer turned survivalist. A follower of Gary North, Olmsted moved to the desert in a mobile home, stockpiled food, and purchased his first gun. An entity titled Comp.software .year-2000 newsgroup launched in November 1996 was said to be ground zero for the Y2K survivalist movement. Survivalists believed that all infrastructure, nuclear power plants, communications, supply chains, government operations, businesses, and law enforcement would fail on January 1, and that civilization would devolve to the law of "only the strong survive" and "anything goes." Steve Watson, an Oklahoma systems analyst, built a forty-person camouflaged bunker, embarking on complex power, provisions, and protection efforts. He and a secret partner purchased 500 remote acres in Oklahoma for their retreat.[7]

In a related article, *Wired* described Gary North and his radical Y2K site garynorth.com, "the oldest, most notorious Y2K doomsday site," as widely read and reported. North predicted that if the electrical grid went down on January 1, it would be permanent due to lack of electricity to fix the Y2K issues. Credible examples of electrical grid failures described "00" misreading of maintenance logs, temperature sensors, or protective devices, causing power plant shutdowns. The North story, for example, ended with an ironic scenario: "Remote areas may remain dark for weeks or months after January 1, 2000, leaving Y2K survivalists waiting in their isolated cabins for the lights to come back on—while complacent urban dwellers enjoy uninterrupted service."[8] More conservative websites like Y2Knews.com continued to push a narrative of Y2K chaos.[9]

Though the majority of people in the United States did not move "off the grid" or prepare anywhere near as drastically as the hardcore survivalists, the media attention given to them did have an effect. Reporting survivalist hysteria in respected publications with numerous reports of readiness failures lent an air of credibility designed specifically to frighten the public.

Robert Huebner's *Public Forum* article, "Y2K's Lessons for e-Commerce," served as an interesting contemporaneous comparison of Y2K with electricity in the early 1900s and predictions of e-commerce. Huebner argues that when electricity access and usage expanded from the business domain into the household, it enabled tremendous social and commercial upheaval–"No one remained unaffected." Huebner went on to predict the same upheaval with e-commerce, which, he wrote, would surely be a business focus following Year 2000 remediation efforts. Very presciently, he stated, "No one is safe to assume that business—and daily life—will ever be the same again." He also cites the U.S. insurance industry (both casualty/property and life and health) as (then) being a $700 billion industry, spending $121 billion on distribution with 20 cents of every premium dollar spent on paper policy distribution.[10]

Boston Herald opinion writer Steven Avellino, citing historic precedence, offered in mid-1999 that it would be the essence of capitalism to protect the nation and the globe from Y2K catastrophes, stability being key. In capitalistic societies, "institutions with the most to lose will protect their interests." Avellino referenced repayment of Revolutionary War debts, the private bailout of the U.S. government during the War of 1812, and three tax rate reductions in the 1920s as examples of historic stabilization; "The stability of government, as a democratic institution, is based on faith. If government computers succumb to Y2K, there will no longer be the perception of stability, and all faith will be lost." Avellino was convinced, however, that capitalism will see the nation through the millennium turn.[11]

The *Boston Globe* quoted Needham computer consultant Steve Goldberg four days before the new millennium as noting "economic self-interest" as the motivating factor for businesses.

"Nobody can afford to take the public-relations hit of a Y2K failure," he said.[12]

In a "Briefs" piece published in the insurance journal *National Underwriter* in mid-1999, Mark Grady, "An expert in law and technology" then chairman of the National Center for Technology & the Law and dean of George Mason University School of Law, made an odd proposal. "Businesses," he declared, "should not be liable for economic losses from the Year 2000 computer problem if the losses are not 'foreseeable.'. . . People cannot avoid committing thoughtless slips and blunders, which are negligent." His argument continued, saying that the magnitude of the problem was so large, insurance would be difficult to obtain; thus, "the courts have to put limits on liability when insurance is very difficult to get."[13] This is worth thinking about on many head-scratching levels.

In a January 1999 article in *National Underwriter*, it was suggested that for the first time, the global nature of this millennium date change posed new issues for insurers. Consider this, the article proposed: a claims-made global program written in the UK, including both U.S. and Australian locations with an expiration date of January 1, 2000; a Y2K failure at 9:00 p.m. EST December 31, 1999 in the United States would extend beyond the 12:01 a.m. 01/01/2000 UK expiration date. Would coverage apply? Reinsurers in particular were not explicit on time jurisdictions. Recommendations were offered to avoid both December 31, 1999 and January 1, 2000 renewal dates, both very common for commercial risks.[14] This example, posed exclusive to the millennium, would actually apply at any time, when a policy expiration date—customarily at 12:01 a.m. on that date—causes difficulty with a foreign loss occurring as above.

Boston's *Mass High Tech*, generally limited to technology issues, ran an editorial in January 1999 that began: "We all know that come the millennium, airplanes will drop from the sky, elevators will plummet to the basement, killing and maiming everyone inside, Wall Street will grind to a halt, communications systems will shut down across the globe, refrigerators will boil milk and washing machines will spin out of control."

Hyperbole, indeed. The editorial ended with, "Indeed predictions of trillions in suits are likely as inflated as those surrounding the future of planes, elevators, and the stock exchange. One thing, however, is certain: Government efforts to meddle by shielding some from exposure while limiting the rights of others is likely to bring only folly to the hysteria that already surrounds the millennium."[15]

"Hoping to cash in on Millennium Bug fever, insurance companies are developing policies to inoculate companies that suffer losses when 1999 flips to 2000 and their computer systems mistakenly read 1900," began a mid-1998 *Wired News* article titled "Insurers Offer Millennium Bug Protection."[16] This is a bewildering statement unsupported by evidence. Not only is "inoculate" a term rarely used in reference to insurance, but "companies suffering losses" generally do not contribute positively to an insurer's bottom line as expressed by "Hoping to cash in."

Other opinions include the following statements:

"[O]ne of Esquire's lesser virtues is Tom Junod's report from Pat Robertson's conference on the digital apocalypse. ('The most important thing is finding hiding places for the Jews,' a fellow visitor tells Junod, 'so that they can get back to Israel, so that Jesus Christ can come again in glory.')"[17]

"I don't think anything is going to happen, other than a somewhat dismal New Year's for the poor people who have to spend it in their offices."[18]

"Not a thing. Whoever believes in that stuff is so gullible it's pathetic."[19]

A widely publicized theory called the "Crouch–Echlin Effect" reported that computer clocks would jump randomly and occasionally either backward or forward at the millennium turn, affecting Intel or Compaq computers when they were turned on. Known as time dilation, tests by computer experts, both

Courtesy of the author.

independent and at Intel and Compaq, refuted Crouch–Echlin, saying "faulty components rather than design bugs were undoubtedly responsible for the reported date jumps." Michael Echlin refused to concede, however.[20]

WIT AND WISDOM (*AGAIN . . . YOU BE THE JUDGE*)

Following the turn of the millennium, the U.S. Air Force embarked upon extensive research growing out of a mid-1998 National Research Council (NRC) project titled "Managing Vulnerabilities Arising from Global Infrastructure Interdependencies: Learning from Y2K." The millennium planning was considered "an extraordinary opportunity to learn . . . how various factors, including current management structures and practices, impact . . . risk that threatens serious damage to information and other critical infrastructures." With the benefit of hindsight, the report noted, "The fact that Y2K did not produce major sustained disruption for the Air Force or other organizations makes it a more valuable source for long-term lessons for operational and strategic management of ICT systems."[21]

Robert Kuttner in his mid-1999 *Boston Globe* column suggested that Microsoft's Bill Gates was the villain in the Year 2000 story, offering that when Windows software was released in the 1990s there was no excuse for not using a four-digit date field, comparing Microsoft to the 1984 introduction of the Macintosh, which automatically transitioned to 2000. Kuttner went on to assert that throughout economic history inferior technology often wins out, either through superior marketing or embedded standard usage. By way of example, he compared the date field issue to the QWERTY keyboard, which has remained the standard long after typewriter jammed keys were no longer an issue. Kuttner also noted the benefit of mandatory government standards, which would have eliminated the hundreds of billions of dollars of millennium remediation cost and concern.[22]

"Y2K Bug Spray" was the brainchild of an insurance claims adjuster, Dan Dyce, and his buddy, Dale Miller, to market a spray bottle said to be 99.746% Y2KBS (or pine-scented water). Sold as the Christmas stocking stuffer of choice for COBOL programmers, a technical meeting icebreaker, or advertising tool, it came with the promise of "being a solution" if used by 01/01/2000.[23]

In *Software Magazine*'s October 1998 final "Year 2000" focus issue, Ian Hayes closes with a tongue-in-cheek look ahead to 2004 for IT professionals: "Remember when you used to just slap changes into programs at will, documentation was for sissies, and testing was throwing the whole mess into production to see if it worked?" "Call it a beta release and let the business users sort it out!" "Those were the days!" Hayes predicted that Y2K litigation as well as subsequent legislation would raise the bar on IT performance standards, documentation, quality, and testing. Noting both liability insurers' higher standards and coverage costs, Hayes concluded that post-millennium programming will be more of a science than an art. The top of his list of "Winners" was lawyers, with Al Gore on the "Losers" lists. "Don't say I didn't warn you!"[24]

On May 9, 1999, *Parade Magazine* columnist Marilyn vos Savant answered an inquiry about Year 2000 concerns with a longer-than-usual reply, declaring "I believe that concern about the Y2K bug will be the first major Internet-driven scare of the global information age, aided by alarming media reports." She went on to itemize a succession of everyday problems, from air-traffic control issues, common red tape problems, to late arrival of mailed checks. "On a scale of 10, my current guess of Y2K consequences ranks between 1 and 2," 1 for inconveniences and 2 for circumstances involving vulnerability, concluding with "other than scaring half of us half to death, any serious consequences will be scattered and only temporary."[25] Having been a vos Savant reader for decades, I consider that prediction of hers most prescient.

ADVERTISING MISADVENTURES

Polaroid Corp., the premier maker of instant cameras before the introduction of mobile phone cameras, created immediate ire from the banking industry when it ran an ad depicting the camera taking a photo of an ATM machine million-dollar windfall right after midnight, January 1, 2000. Banking lobbyists suggested that Polaroid would not want banking ads depicting an inoperative Polaroid Land camera. Polaroid's defense offered that the company was trying to attract younger buyers, combining humor with contemporary events.

Polaroid was not the only advertiser to use banking problems in their ads; Kia and Volvo were doing the same, according to a *Boston Herald* article mid-1999. Complaints to the American Association of Advertising Agencies was met with the response that the AAAA did not direct members on what and what not to include in advertising material.[26]

Courtesy of the author.

ADVICE ON DEALING WITH THE PRESS

William McDonough of Marsh, Inc. (Boston), offered recommendations on dealing with reporters in an article specific to hospital Y2K readiness. In addition to advising that "no comment" is completely unacceptable, McDonough offered the following tips: update reporters frequently, always stay on the record, cooperate with the media, talk in plain English not jargon, do not play down the crisis, tell the truth, and be preemptive if possible. He stated that he knew of no examples when the media has not gotten to the truth when questions were evaded or misleading replies given.[27] I am including this because, in the thousands of Y2K documents in my files, almost none deal with advice on media and press inquiries.

COMPUTER PROBLEMS PRIOR TO 01/01/2000

Software problems had been causing difficulties decades before Y2K. Among the most tragic occurred in 1991, where "a Patriot missile defense system operating at Dhahran, Saudi Arabia, during Operation Desert Storm failed to track and intercept an incoming Scud. This Scud subsequently hit an Army barracks."[28] The error was described by consultant Jon Huntress as an "unrecognized clock drift" bug in Patriot air-defense batteries resulting in a 678-meter tracking error.[29] That tracking error led to the deaths of twenty-eight National Guard service personnel hit by an unintercepted Scud missile. Referencing the Gulf War tragedy, Deputy Defense Secretary John J. Hamre testified before the U.S. Senate: "If we built houses the way we build software, the first woodpecker to come along would destroy civilization."[30] Tragically prophetic.

Other significant software troubles: A programming error in an F-16 fighter plane simulator would cause the plane to flip upside down whenever it crossed the equator;[31] a Year 2000 test by the Chrysler Corporation resulted in a time clock malfunction, causing a security system shutdown that kept everyone from leaving the building; and a Philips Petroleum Y2K simulation on a North Sea oil vessel caused the failure of a safety system designed to detect deadly hydrogen sulfide gas.[32]

SCAMS

Fraud schemes were a natural part of Year 2000. While I have noted that rogue states did not have a major role in attacking their enemies with Y2K havoc, thieves looking for financial gain were proliferating. Callers seeking bank account information with stories about moving depositors' accounts into special "bond" accounts for Y2K testing seemed credible, as was the same tactic seeking credit card numbers and data.[33]

Reports of criminals taking advantage of Year 2000 disruptions during computer remediation upgrades were said to be scamming billions of dollars from government agencies and private companies. Creating false names, addresses, social security numbers, and identifiers, thieves filed for false insurance, Medicaid, and Workers Compensation claims. With the technology distractions of Y2K, insurers missed issuance of duplicate policies, resulting in one case of $100 million stolen by accident-staging fraud rings from a casualty/property insurer.[34]

Fraud rings were skilled at knowing what technology shortcomings specific insurance companies had; one fraud ring moved from region to region to make false Workers Compensation claims against one insurer without detection due to the absence of nationalized claim data. Computer-savvy fraud families, multiple name and location spellings, and large complex rings recruiting undocumented immigrants took advantage of the Year 2000 opportunities.[35]

By late 1999, Y2K viruses were being launched by "hackers and unscrupulous hooligans," according to Don Jones, Microsoft's director of Year 2000 readiness. One such 1999 virus, using the Windows operating system, sent e-mail messages to Outlook addresses with the message "Here's some pictures for you." The attachment pics4you.exe, when opened, sent victims to a pornographic website, next erasing data after January 1, 2000. There were a number of viruses disguised to look like Y2K repair programs as well, activating problems on January 3 and 4 when offices reopened.[36] Computer usage protocols for employees became an important priority in avoiding such viruses. Over the next twenty years those protocols have become, out of necessity, more and more complex, adding legislated requirements for privacy and security.

In a *Los Angeles Times* article unrelated to Y2K matters, the National Security Agency launched a team of thirty computer specialists named the "Red Team" to try to hack into government agencies. Electrical grids, 911 emergency systems, and the Pentagon's National Military Command Center were successfully hacked. The FBI's "Moonlight Maze" found that Russian hackers had been downloading an impressive array of Defense Department research data. These vulnerabilities were discovered using the simple process of buying laptops, downloading software, and targeting unclassified computer systems.[37]

THE YEAR2000 POLICE BOTHER YOU AT LUNCH

Courtesy of the author.

NOTABLE QUOTES

> I predict that the Y2K explosion will blow up Mr. Gore's political bridge. The odds are higher that he will be swimming in the River Kwai than sitting in the Oval Office in 2001.—Ed Yardini, chief economist, Deutsche Morgan Grenfell[38]

> Computers nowadays do their job so well, and are so transparently efficient, that only the programmers who have written the software know what a Rube Goldberg labyrinth lies just beneath the smooth tin skin of these machines.—Jon Huntress, consultant for contingency and disaster recovery planning[39]

> *Scientific American* magazine compares the current state of software development to the state of industrial development before Eli Whitney invented the assembly line in 1799. In other words, software is essentially a handicraft item.—Jon Huntress, consultant for contingency and disaster recovery planning[40]

> First there was mad cow disease, now El Nino. The whole Y2K thing is just another in a series of catastrophes. Before I came here I thought all the doom and gloom was too much. I want to make sure I get my paycheck, things like that. If the power goes out and I am in Canada, it's going to be cold.—Paul Mosher, product marketing manager of Mitel Corp., Ottawa, Canada, at the November 1997 "The Year 2000 Conference & Expo," Boston[41]

Accounting arbitrage was referenced in a respected online discussion group of @year2000.com. An inquiry from Patrick Vice at @ibm.net from ADVice of Toronto, Canada regarding "Insurance for Year 2000" was answered by amy@year2000.com, a repeat contributor to that chat column. "The bigger insurance brokers are up to speed on Year 2000 and are offering policies (at least in my industry which is Banking.) Some are accounting arbitrage vehicles (basically a bookkeeping trick to let you write off more of the project cost with a 15% 'coverage' of traditional insurance against unforeseen costs) while others are offering insurance against vendor failures or missed deadlines."[42] Such a reply on a prominent Year 2000 website left me bewildered, both as to the actual product(s) offered as well as the source of the information. I knew of no such insurance or quasi-insurance vehicles to have been offered, even while understanding there were accounting cost-allocation issues.

John Dodge, *Boston Globe* correspondent, related running Y2K tests on his personal computer. Y2000RTC and Know2000 were described, with response regarding his Dell PC's Windows 98 compliance. Dodge noted that this process illuminated the number of obscure applications, like Twain Thunker, that he had never known existed—all part of Microsoft packages.[43]

In Robert Sam Anson's January 1999 *Vanity Fair* feature article, he commented that "Russia has 11 Chernobyl-type power plants, 22,500 nuclear warheads, and the funds to fix none of them."[44]

Church bells all over Britain were preparing to toll for five minutes to ring in the new millennium. More than 6,000 churches possessed sets of five or more bronze bells, which required skilled ringers. Britain admitted needing 5,000 additional bell ringers before December 31, noting that "snaking bell ropes can launch an unwitting greenhorn toward the ceiling."[45]

GAFFES

The U.S. Navy withdrew its report that sewer systems were not Y2K ready in naval base towns of Quincy, Massachusetts; Newport, Rhode Island; Kittery, Maine; Groton, Connecticut; and Philadelphia, among others. Navy Commander Hank Chase said "those findings reflect incorrect formatting on the website." This follows an August "backpedaling" of the Navy's report of utility readiness.[46]

In a mid-1998 New Jersey Online article, it was reported that property/casualty insurer Reliance Group Holdings of New York was "involved in the software business through RCG Information Technology in Edison [New Jersey]." The fairly lengthy article went on to describe the operational and financial details of RCG, but nowhere mentions that it, or Reliance, was offering a Year 2000 insurance product.[47] The regulatory and filing complexities alone would be prohibitive for anyone but an established insurer to offer any insurance product, let alone a technology-based enterprise.

RUMOR CENTRAL

CPSR (Computer Professionals for Social Responsibility), founded in 1981, is a global organization promoting the responsible use of computer technology. In mid-1998 CPSR established Rumor Central to help sort Year 2000 fact from fiction. Helpful categories of Rumor, Speculation, Prediction, Unconfirmed, Fit To Print, and Contradicted, displayed in grid format most of the issues that headlined Y2K news stories. This book discusses those stories; Rumor Central's categories helped readers to determine the relative weight of rumor, unconfirmed, and speculation versus predictions from experts and reliable sources.[48]

Selected "Rumor" from CPSR Rumor Central:

1. Nuclear weapons will be accidentally launched (unattributed).
2. More than 30 percent of companies will be exposed to a millennium bomb threat (unattributed).
3. The United States has already lost a few satellites with Y2K glitches (unattributed).
4. The date-dependent IRS could implode.[49]

Among the list of more interesting "Predictions" and "Speculations" are:

1. There will be a 100 percent chance of a worldwide recession on the order of an OPEC-sized event as in the 1970s.[50]
2. Insurance claims will go unpaid to more than 1 million people.[51]
3. Land mines will explode (unattributed).
4. The U.S. Department of Defense will shut down.[52]
5. Building security systems won't let anyone in (unattributed).
6. GPS receivers will fail.[53]

These eclectic quotes from the time exemplify the variety of human condition and opinion. This does not seem to vary, generation to generation, crisis to crisis.

THE
YEAR2000
POLICE
WAKE YOU
AT 3:00 AM
TO REMIND
YOU OF
SOME
DEADLINE

Courtesy of the author.

Technology Insurance and Risk Specialists

Call, Fax, or E-mail today
(508) 369-2771
Fax (508) 369-2778
104377.2437@compuserve.com
www.npjames.com

Y2K Police postcards advertising my services and contact information. *Courtesy of the author.*

10

CONCORD, MASSACHUSETTS, AND YEAR 2000 READINESS

In the lead-up to the new millennium, I knew I needed to take an active role in Y2K readiness, primarily for issues of public safety. In mid-1997, I addressed the American Bar Association at their annual meeting in San Francisco, focusing on Y2K liabilities not only in the corporate world, but in government as well. In June 1998, I flew to London to address EU and U.S. insurance industry leaders on the same topic.

After returning to my home in Concord, Massachusetts, I happened to run into the Concord Town Manager, Christopher Whelan. When I inquired about Concord's Y2K initiatives, Chris told me that the town's computers were being reviewed for compliance. I offered wryly that late tax notices delayed by noncompliant computers were not an issue I cared deeply about, but the reliability of our town's heat, power, and communications grids in mid-winter certainly were. Chris acknowledged that he was unaware of those issues.

I'd lived and worked in Concord, Massachusetts, for over twenty years in 1998, and I live there to this day. Located northwest of Boston, this Massachusetts town is one of the birthplaces of U.S. history, drawing visitors every year to visit the scene of the "shot heard 'round the world" and other famous sites pivotal in the American Revolutionary War. It also hosts the homes of famous authors such as Ralph Waldo Emerson, Louisa May Alcott, Nathanial Hawthorne—and of course, it is also the home of Henry David Thoreau's Walden Pond.

It was important to me that my own town was ready and informed about Y2K. After my conversation with Chris Whalen, I resolved to get to work locally. This chapter details the efforts that I and other residents of Concord took to address the wider implications of computer problems heading into the new millennium.

In subsequent Y2K issues conversations with Mary Johnson, the executive director of the Concord Chamber of Commerce, I was asked to conduct a Y2K forum for the community. Mary arranged the venue under the Chamber's auspices while I contacted speakers to discuss the municipal issues facing us with the millennium date rollover. The first, Forum 1, was scheduled for January 22, 1999, followed by Forum 2 on September 23, 1999.

In initiating a Y2K readiness information forum, my primary priority was Concord's public safety. Gathered at Newbury Court was an impressive group of public safety officers and state and local officials. Supported by the Concord Chamber, I convened Christopher Whelan as well as Concord Police Chief Leonard Wetherbee, Concord Municipal Light Plant Superintendent Dan Sack, Massachusetts State Representative Pam Resor, and Chris Doherty, Y2K regional issue aide from Congressman Marty Meehan's office. I stated, "This will be a hard look at the town officials' projections in early 1999 of the impact of Year 2000 computer problems to infrastructure, public safety and other matters." With experience and knowledge in both insurance and associated liability law, I opened the forum by offering perspective on exposures and litigation surrounding Y2K. Both State Representative Pam Resor and I addressed the state's emergency plans and possible legislation related to liability.

What follows is the text of *Concord Journal* articles before and after Forum 1.

January 1999 Year 2000 Forum Announcement
Press Release

The Concord Chamber of Commerce has invited Year 2000 issues expert, author, national and international speaker, Nancy James, owner of N. P. James Insurance Agency, to structure and moderate a panel on the community (public) safety, infrastructure, and legislative issues of the anticipated Millennium computer failures. Speaking on the panel, to be held 7:30 AM January 22, 1999, at Newbury Court's main dining room is Nancy James, Moderator, Chris Whelan, Town Manager, Len Wetherbee, Concord Police Chief, Dan Sack, Concord Municipal Light Plant, Representative Pamela Resor, and Chris Doherty, Congressman Marty Meehan's Year 2000 representative. Interested attendees should call the Concord Chamber of Commerce 978-369-3120. Cost is $10.00 for continental breakfast.

James has addressed law and insurance audiences in the United States and Europe on the liabilities associated with Year 2000 failures.

JANUARY 22, 1999, YEAR 2000 FORUM

Concord's local cable television network CCTV filmed my January 22, 1999, forum for broadcast. Opening the forum with short introductory remarks, I first thanked Mary Johnson and the Chamber of Commerce. Setting the purpose of the program as examining community safety and infrastructure issues, I referenced the global nature of Year 2000 computer remediation issues, as well as the rather grim readiness statistics published by national expert Capers Jones. The *Concord Journal*'s report of the Forum follows, but I will add those highlights I think are important to the story from my re-viewing of the CCTV video.

Town Manager Chris Whelan led the distinguished panel's remarks. He referenced the town's 15,000 residents, $50 million budget, and 500 employees, as well as reassured the public that water systems contained manual overrides for continuous service and a $150,000 set-aside for problems such as shelters and schools. Referencing my earlier comment to him about not minding late tax bills, Chris assured everyone that tax invoices would be delivered on time.

Police Chief Len Wetherbee followed, opening with information that the police department had purchased a new records management system in 1994 that was Y2K compliant, which included the police dispatch system. The long-standing E911 system, however, was Y2K noncompliant, which Bell Atlantic was not going to address until the second half of 1999. Concord's old 1212 emergency number had not been disconnected and would still be useable in the event 911 was not. Len continued by describing how Route 2 highway workarounds would be set up in the event traffic lights were inoperable on January 1, stating that the highway protocol would be the same as if it were a natural disaster. He anticipated the station would get 200–300 false alarm calls from residential noncompliant alarms.

Concord Municipal Light Plant Superintendent Dan Sack was third, advising those gathered that there were 4.6 trillion lines of electrical code in the United States. He described the New England Power Pool (NEPOOL) connecting all of New England, as well as the power grid connections from Canada to Florida and how another 1965 northeast blackout would currently be avoided. Manual operation of Concord's power station would be initiated for a local Y2K failure issue, but if power could not get to Concord, there was little that could be done.

State Representative Pamela Resor followed, giving us the Massachusetts Y2K readiness temperature, indicating a 70 percent readiness, and pointing out that Year 2000 employment has resulted in a boost to the Massachusetts economy. From $70 to $100 million was being spent on compliance, and a Y2K committee had been formed in June 1997. Massachusetts, like other state and federal government entities, had filed legislation to hold harmless government entities and hospitals . . . *just in case. . .*

U.S. Congressman Marty Meehan's Year 2000 representative, Chris Doherty, was the last speaker on the panel. He mentioned that President Bill Clinton's State of the Union address included Year 2000 as a top priority, and in February 1997 the president had established a Y2K Council. Doherty offered the national telephone number, 1-888-USA-4Y2K for people who had Y2K questions and concerns.

The Q&A portion followed with some interesting and thoughtful questions. Among the questions, Valerie Kinkade asked Chief Wetherbee about martial law. Len replied that he had his twenty years in! (jokingly referring to his immediate retirement in the event of martial law). Len then offered that the blizzard of '78 came close, explaining that martial law meant that control of public safety would be taken over by the military. Another audience member inquired about looters in the event of Y2K infrastructure collapses. Len replied that it would not take long to overwhelm the town's thirty-four officers, at which time the National Guard (martial law) would be called in. A question regarding every traffic light needing a chip replacement was again answered by Chief Wetherbee, who replied that Concord was not equipped to provide two uniformed officers at every intersection, describing how barrels would be set up at every Route 2 light to weave traffic into and through Concord to the next intersection.

Mary Johnson concluded the program by thanking the speakers and audience on behalf of the Concord Chamber of Commerce.[1]

January 22, 1999, Y2K Forum—*Concord Journal* article

In the January 14, 1999, issue of the *Concord Journal*, the Forum was announced, interviewing Chris Doherty, Congressman Marty Meehan's Y2K regional issue aide:

Chris Doherty, Y2K regional issue aide from Congressman Marty Meehan's office, said that Concord is one of the first communities in the state to organize a town-wide discussion of the Y2K problem. "It's important from a Congressional standpoint because it's a major national issue and it's an issue a lot of local towns are taking seriously. Forums are really something that's just beginning and Concord is really on the cutting edge. This forum in Concord is a great education opportunity for all parties involved. We can come together and have a great discussion on this issue."

Doherty offered that he would bring a federal government perspective to the Forum, including updates on Y2K spending, successful remediation efforts, and public education programs. A 24-hour toll-free hotline to provide information about the Y2K problem has been set up; 1-888-USA4Y2K.[2]

Concord Journal reporting on the January 22, 1999 Forum in its January 28, 1999 issue

Organized and led by Nancy James, a technology risk manager, she introduced the forum by stating that the United States was a world leader in the $600 billion global effort for Year 2000 computer system remediation. However, she offered that there was no state or federal legislation to ensure that computer-controlled mechanism like bank ATMs would successfully make the switch from 1999 to 2000. James added that "Americans love solving problems at the 11th hour and that European nations are looking to the United States for answers."

Concord Town Manager, Chris Whelan explained that all new software the town buys is Y2K compliant and that remediation of all older existing software is underway. "The town treasurer's office has $60 million at stake in tax receipts. You can bet they take the problem seriously." As of January 1999, 70% of the town's PCs were compliant as efforts continue. While he suggested there was no need to panic, Whelan also added that there was no reserve fund for remediation efforts.

Whelan added that a manual override of the Concord Public Works Department water delivery system is in place should automated systems fail.

Police Chief Len Wetherbee offered comments on public safety, stating that communications systems would be compliant by year's end, including the 911 emergency number. He went on to outline the standard protocol if Route 2 traffic lights went out, and that the department would be staffing for a major natural disaster on 12/31/1999.

Dan Sack, director of the Concord Municipal Light Plant commented on the efforts of the New England Power Pool (NEPOOL) with a $70 million budget for remediation, including reprogramming for generators and reviewing and updating of 4.6 trillion lines of code. He stated that additional staff would be at "power stations to override non-compliant systems and keep things running."

Sack added the electrical grids serving New England, New Jersey, New York, and Connecticut, while interconnected, New York and New England were isolated following the massive 1965 power outage.

Chris Doherty, Y2K regional issue aide from Congressman Marty Meehan's office, offered U.S. Senate and House websites for the public to follow Year 2000 efforts, as well as a national hot-line number citing the President's Year 2000 Conversion Council, consisting of more than 30 federal agencies.

State Representative Pam Resor reported that Massachusetts had spent $70 to $100 million on compliance with agencies providing quarterly reporting to the legislature. She added that Massachusetts leads the six New England states in progress to date, but was last with respect to contingency plans, offering the public can follow progress via the www.state.ma.us/Y2K website.

The *Arlington Advocate* newspaper in a February 18, 1999 editorial

There is hope on this somewhat cloudy horizon. A recent forum in Concord brought optimism that local officials and utility representatives are well ahead of the problem and are looking ahead to the Year 2000 with their ducks in a row. Unfortunately, not every community, state, country or company can say the same.[3]

Valerie Kinkade became a valuable voice in Concord's preparations, offering indispensable perspective and thoughts to disaster education and planning. Attending my January 22, 1999, Forum, Valerie raised some interesting areas, including her singular experience in a California earthquake shelter, offering vividly how shelter living was not ideal. Her letter to Concord Municipal Light

Plant head, Dan Sack, raised questions about the reliability of our 911 emergency number, safe food and water storage, fires and fire detection systems, and determining what businesses had backup generators. Her suggestion of using polling places for information dissemination should communications systems be down was particularly noteworthy.

Valerie Kinkade then became a founder of Concord Neighborhood Network, bringing to light that government agencies' Year 2000 training sessions were not open to community attendance. A copy of Gary North's Y2K Links and Forums, titled "FEMA Holds Closed Conferences. Were You Invited?" was forwarded to me, advertising a mid-February 1999 national Y2K Preparations & Consequences Management Workshops conference in Atlanta for federal and state government officials. Regional workshops, including one in Boston, were listed as well. "Attendance by invitation only" was highlighted in large, bold, italic type.[4] Valerie had asked Joe Lenox, Concord's fire chief and head of Civil Defense, if citizens might attend some of the local training sessions to better assist in community preparations and response. She was understandably disappointed by Joe's rather perfunctory reply refusing her request with a note that "Input from concord [*sic*] citizens is always welcome."[5] While Joe Lenox was in general extremely cooperative and forthcoming to citizen inquiries and activists, government jurisdictional lines were clearly in place.

Approaching my September 1999 forum, our community faced more citizen uncertainty and anxiety. In my mid-August letter to energy consultant Richard D'Aquanni, I explained that I had alerted my panelists to be prepared; the upcoming forum might be less cordial than January's. "As public awareness grows, so do public concerns. I have been interested to observe attendees at some recent Y2K community meetings. Skeptics scoff at the data given by presenters. . . . So public doubt is growing too, while I continue to maintain that this issue enjoys the greatest discrepancies in opinion and data ever seen by our culture! I have some foreboding about our 9/23 Forum!"[6]

SEPTEMBER 23, 1999, YEAR 2000 FORUM 2

The second Y2K Forum for Concord, Massachusetts, on September 23, 1999, was held at Concord's town hall, known by residents as our Town House. It was important to capture at this eleventh hour the honest assessment of computer readiness for the millennium change. By this time both my work and concerns were being noticed by state and town officials concentrating efforts on solutions.

Forum 2 panelists were State Senator Susan Fargo, Concord Town Manager Christopher Whelan, Concord Police Chief Leonard Wetherbee, Concord Municipal Light Plant Superintendent Dan Sack, and Concord Fire Chief and Civil Defense Head Joe Lenox. Each offered information on infrastructure readiness, civil defense plans, fire emergency safety, as well as Commonwealth and global compliance efforts. Questions and concerns from the audience were addressed. While I was moderating the event, the *Concord Journal* staff writer Greg Turner was taking notes for his newspaper article (the quotes he captured are cited in my endnotes).

State Senator Fargo reported that the Massachusetts' Legislature's Science and Technology Committee was holding hearings during the year for readiness review of utility companies, hospitals, and state agencies. She had sponsored a February 4, 1999, State House Forum for 100 local officials and computer experts.[7] Offering confidence in the Commonwealth's readiness efforts, Senator Fargo added that the Massachusetts Emergency Management Agency (MEMA) had contingency plans ready for January 2000 in the event of issues.

Concord Municipal Light Plant Superintendent Dan Sack reported that as of the date of this September forum, 99 percent of the nation's mission-critical electrical systems were compliant as reported by the North American Electric Reliability Council, the nonprofit agency overseeing Y2K for the U.S. Department of Energy. Sack added that Concord's electrical grid systems had been successfully checked for compliance and that he was confident that the lights would stay on as Concord celebrated the new millennium.

Police Chief Len Wetherbee followed by saying, "My job is to focus on everything that happens if Dan's wrong. Hopefully he's right and we won't have any problems."[8] Wetherbee continued, noting the disruption when alarms are triggered by electrical outages, as well as possible disturbances of the new millennium's New Year's Eve celebrations. Police resources would be bolstered by the Northeast Massachusetts Law Enforcement Council's twenty-six-town consortium, since Wetherbee expected "the largest parties in Massachusetts that we've ever seen in our lifetimes."[9]

Interestingly, those huge parties did *not* seem to materialize in the Boston area, as many technology professionals were working New Year's Eve, reportedly filling all of Boston's hotel rooms in anticipation of financial sector Y2K computer issues.

Civil Defense Head Joe Lenox recommended that residents and businesses should prepare for Year 2000 in the same manner as a major hurricane or snowstorm. Lenox noted that intermunicipal agreements for emergency resources would likely not be available in the event of a statewide Y2K power outage; "We're going to be on our own."[10] However, Concord has prepared shelters and has a list of residents requiring reliable electrical service for ventilation or dialysis.

Hurricane Floyd hit New England on September 16 and 17, 1999, offering town officials a ready test for Y2K preparedness for power outages shortly before our Forum 2 on September 23. "We were thrilled we got a chance to test [our readiness]," said Charleen Sotolongo, manager of safety and security at Emerson Hospital and member of the Emergency Planning Committee. Sotolongo stated that existing statutes require Emerson Hospital to have backup generators. "Any critical life-support systems will continue to run and will not jeopardize the care of our patients."[11]

A very important point was raised by an attendee, who noted that Concord at one time had a radio station. He suggested that radio communication could be vital to broadcast information and would avoid clogging police telephone lines in the event of electrical or other communications disruption. As a result, the Town's renewal of the radio station license was made just days before it was to expire. Shortwave radio training was offered to volunteers for Y2K. The station remains in operation to this day; again, only because it was saved (by Y2K!) just days before expiring. Unintended benefits!

Another attendee asked the Concord Municipal Light Plant Superintendent what would happen if Canada's power grid went down and Canada requisitioned all of Hydro-Québec's power (since at the time Concord purchased electrical power from Hydro-Québec). Answers to such larger-picture questions were mostly still unknown, as local officials in Concord and everywhere struggled to identify and control infrastructure and services.

When Police Chief Len Wetherbee was asked if, in the event of a power failure, the MCI Concord prison gates would lock open or closed, he smiled confidently and replied "closed." Following the forum, however, Len mentioned to me that he thought he might need to make a call when he got back to the station.

The forums offered the community a lot of value. They offered state and local officials the opportunity to gather and hear each other's operational functions and progress. Citizen input and concerns could be heard and regarded as valuable perspectives. The forum was overall cordial, but the anxiety in the hall was definitely palpable; residents were understandably uneasy about the Y2K magnitude and complexity for a successful outcome.

In contrast, newspaper reports of community readiness described very little in the way of citizen input in the town of Acton, Concord's neighbor, with Y2K expert Capers Jones of Acton stating, "I am least optimistic about city government. We have not done a good job." Should basic infrastructure services fail due to Year 2000 issues, lawsuits against their towns from local citizens would be likely, said Jones.[12] Experts like Jones were assisting American cities and towns with both preparations and worst-case contingency operational planning. Referencing both the possibility of lawsuits, litigation offense and defense, as well as political fallout if Year 2000 remediation failures caused issues with municipalities, Jones's abstract outlined emergency protections of concerns like banks, public facilities, and food sources. Communications, heat, power, shelter, needs for federal protection intervention, and infrastructure protection were part of past 01/01/2000 worst-case contingency planning.[13]

As a consequence of the planning of the two forums I organized and conducted in 1999, the Concord Emergency Management Agency (CEMA) began holding monthly meetings for Year 2000 emergency preparations. Members included the Select Board (named Board of Selectmen at the time), town manager's office, fire, police, public works, light plant, Local Emergency Planning Committee (LEPC), and school maintenance and transportation departments. CEMA prepared for a two-week utility interruption, with each of the community segments preparing emergency services for their constituents.

Concord operated its own municipal light plant, purchasing and providing electricity to residents and businesses. The head of Concord Municipal Light Plant (CMLP) at the time was Dan Sack, a well-respected manager responsive to both community power issues and to me during our planning of the forums. To advise residents about the reliability of our electrical grid, Dan began sending me documents that he was receiving from NEPOOL (New England Power Pool) regarding software date issues. NEPOOL and ISO New England issued an immediate notice on August 20, 1999, regarding a Bently Nevada Corporation's monitoring software fault that would reset clocks the following day—August 21,1999 to August 22, 1999, to read January 1, 1980. Full backups were recommended for August 20 as well as December 31, 1999 to prepare for system failure during "week number rollover" and "year number rollover."[14]

This specific issue demonstrates an example of the myriad software problems, large and small, constantly being handled by municipal and regional authorities charged with keeping our lights on every day. Another email to Massachusetts, Maine, and New Hampshire regional and interstate utility consortiums forwarded from Dan announced, "September 9 Y2K Drill Sessions" training locations, dates, and times in advance of a Year 2000 drill.

CMLP issued a "Town of Concord Year 2000 Statement" in their *LightLines* newsletter of July–August 1999 indicating that the town's departments were confident about Year 2000 readiness. Light Plant equipment was primarily electromechanical, not subject to computer date disruptions. CMLP made a statement that computer-dependent equipment had been checked for compliance and corrective measures either taken or were in process.

In addition, a local Concord community network project was organized to work with neighborhoods for information and planning. Specific to the network efforts were meetings to help neighbors identify risks, resources, health hazards, special needs neighbors, and sources of assistance. Many community organizations were networking with one another as well as centralized bodies like the Center for Y2K and Society, which asked community groups to rate local preparedness.[15]

While news of local forums reported positive Y2K progress among experts and attendees, there were reports of frustration in other parts of the country. A discouraging late-1999 *Sunday New York Times* article about the national gathering of local Y2K activists in Boulder, Colorado, indicated weariness over the apathy toward Year 2000 preparedness. A speaker from Santa Fe, New Mexico, addressed an audience of only thirty participants for the local Chamber of Commerce Y2K meeting from an invited 1,100 businesses. Neighborhood-to-neighborhood preparedness in Santa Rosa was reported to have collapsed. I'm including this as a rare national report of grassroots efforts to educate local citizens about basic Year 2000 preparedness. This group had previously only been in contact via email with one another. Reportedly "thrown together on a shoestring," Boulder was chosen because of its vigorous Year 2000 preparedness advocacy, described by Kathy

Garcia, head of a Boulder County Year 2000 group, as "27 square miles surrounded by reality."

Poking fun at Y2K, the *Times* article also parodied stanza 1 of the Notre Dame fight song:

> Three cheers for Olde Y2K,
> Stock up on Twinkies without delay.
> Get some water and firewood too,
> We'll spend a fortune before we are through.

The Boulder preparedness evening concluded with flashlights in the dark, a simulated campfire, and sharing of Year 2000 horror stories.[16] Charming and amusing levity.

Just thirteen days before the millennium date change, a page 1 article in the *Boston Sunday Globe* shared interviews with suburban town officials regarding local preparedness. The article also featured neighborhood organizations, including the Concord Neighborhood Network of 112 neighborhood groups and 150 contact names to collaborate on local emergency services as well as identify those with special needs within neighborhoods. Local residents mentioned that Y2K preparation efforts brought back memories of the storm of 1978, where neighbors pulled together to help neighbors.[17] Sales of generators were particularly brisk, and Scott Vanderhoof of Concord's Vanderhoof Hardware noted a run on battery-operated table lanterns. (I still own one!)

To add a local note, the same *Boston Globe* article cited the director of safety and security for Concord's Emerson Hospital, Charleen Sotolongo, as stating that the hospital had conducted a contingency planning exercise during Hurricane Floyd in mid-September 1999, noting only a few medical devices were still awaiting the arrival of software upgrades.[18] Charlene was present at my September 23, 1999, Y2K Forum to offer Emerson Hospital readiness remarks as well as ongoing availability to me and Concord officials as the new millennium approached.

Local Massachusetts town officials' attitudes toward the risks of the Year 2000 date change varied widely. Most local officials described a variety of preparatory measures, with Stoneham's town administrator, Tom Nutting, stating, "I think this could be the biggest nonevent in history."[19] As many Y2K remediators, activists, and watchers remarked as January 1 approached, Y2K anxiety might easily lead to overreaction, which alone could cause problems if everyone picked up the telephone at 12:01 a.m. or dialed 911 to see if it was still working.

And there we were as 1999 drew to a close; cooperating with each other, supporting our local officials in their efforts, and working hard to prepare personally and as a community.

11

AFTER 01/01/2000

20/20 Hindsight

With all of the hype, the fears, experts, expenses, critics, and gainsayers, it is interesting to note the aftermath of what was generally considered a very quiet New Year's Day, January 1, 2000.

While the legal cases involving Year 2000 suits were carefully watched and reported, most other issues seemed to be largely ignored as past news. Y2K practice sections of law firms turned back to their more traditional legal foci. Since information technology departments had, out of necessity, been reassigned to Year 2000 compliance, data security and privacy needed urgent attention and recommitment in the new millennium. I had noted then that while Y2K was absorbing considerable IT resources, hackers were advancing full throttle with more sophisticated tools for breaching corporate and personal data. Legislative bodies directed more scrutiny at the growing issue of privacy protection, enacting more stringent state and federal laws to protect personal identifiable information (PII).

But there are some after-notes regarding Y2K and some interesting occurrences as the first decade of 2000 progressed.

Boston Globe reporter Peter Howe noted in his January 2, 2000, analysis that "many predictions of computer meltdown, driven by the shift to the date 01-01-00 could now be widely disparaged as Chicken Little nonsense. They may even be seen as a conspiracy perpetrated by self-interested vendors of technology."[1] Howe went on to enumerate the billions of dollars spent on compliance and the monumental effort to identify and recode hundreds of millions of lines of computer software code, concluding that the inevitable problems arising after 01/01/2000 would likely be small, likened to "death by a thousand paper cuts" in nature.[2] Of course, with computers shut down until Monday morning, January 3, those were unknowns on Sunday, January 2, 2000, when the article ran.

Howe reported that "Project Magellan" of DC Corp. (Framingham, MA), monitored Y2K problems around the world, estimating a cost globally of $21 billion in 2000, noting that it was only 1/1500 of the $32 trillion total global economy.[3]

Howe also noted that John Koskinen, President Clinton's Y2K Czar, hailed the success of industry and governments in handling "the biggest management challenge the world has had in 50 years."[4] Gary Beach, publisher of the 135,000-circulation *CIO* magazine, declared: "This is one of the best examples of human cooperation in the history of mankind."[5]

All of this was reflective of both great relief and the inevitable euphoria in the few hours after the clocks turned over on January 1. There was more to come. There was also more to be revealed.

THE IMMEDIATE AFTERMATH

Y2K wasn't unique in just its core technical issue—how the aftermath and even its failures played out was very specific to its time. I mentioned in the introduction the absence of cyber warfare—rogue states watching for strike opportunities for attacking or sabotaging their adversaries. This came to mind when reading of a Y2K outage in the Pentagon's reconnaissance satellites at 7:00 p.m. on December 31, 1999 (midnight 01/01/2000 GMT), causing the loss of all, including top-secret, spy satellite highest priority data for several hours. There was a brief television report from the Virginia headquarters of the National Reconnaissance Office early on New Year's Day, then no further details were televised. That report was neither repeated further on January 1 or after, nor was I able to find specifics of it anywhere, even though such a loss of intelligence arguably should have been a top news story. Both the Associated Press and the *Chicago Tribune* ran brief summaries of the outage from the Saturday (01/01/2000) news conference by Deputy Secretary of Defense John Hamre. Continuing in the AP article, Joint Chiefs of Staff head of Y2K preparedness, Rear Admiral Robert F. Willard, had not mentioned the problem at his 9:30 p.m. news conference. It was later learned that Willard had not been notified of the outage by the Pentagon.

The outage lasted for three days, with the most sensitive reconnaissance data being redirected to other sites for manual deciphering. The *Chicago Tribune* reported on January 4 that Pentagon officials were either unwilling or unable to explain how a Y2K problem could occur in an area as sensitive as the nation's spy satellite system.[6]

The *New York Times* reported this on January 4, 2000, toward the end of an article on the inside pages of the Business Day section: "Perhaps the most widely reported problem over the weekend—the failure Friday night of a ground station that processes information from military satellites—was put to rest when the Pentagon reported it was operating normally today."[7] Troubling with the *Times* report is that I had done an exhaustive search then for further information on the early news report, finding none, and that "today" was January 4, 2000, four days after the 01/01/2000 system failure and, importantly, following its restoration. Though we are still operating with very limited information, such an outage (and the lack, it seemed, of adversaries taking advantage) can seem unbelievable to a modern audience. Had sophisticated and far-reaching cyber warfare techniques been available in 2000, the outage could have resulted in catastrophic consequences for the United States.

A mid-1998 *Boston Globe* article reported that the United States was sharing early-warning satellite data with both Russia and China to prevent uncertainty and panic over the possibility of an unknown attack if Y2K darkened those countries. Bruce Blair, nuclear weapons specialist at the Brookings Institution, interviewed Russian nuclear officers, reporting that Russia was relying far more on nuclear weapons following the collapse of its army. However, shortages in both labor and money had weakened Russia's nuclear weapons systems with vulnerabilities. The article also reported top Pentagon officials only learning of U.S. Year 2000 issues in 1995 from the Social Security Administration, then shifting $1.9 billion from weapons programs to Y2K.[8]

THE MONTHS AFTER

Early January 2000 reports of Y2K computer glitches indicated few major disruptions. Nevertheless, there were a few serious issues reported; although likely very few corporate problems being publicized. Like any awkward business hiccup, publication of Year 2000 glitches and interruptions pre– and post–January 1, 2000, were kept as quiet as possible. Michael Hyatt, in a July 1999 e-letter chronicled known Year 2000 glitches, noting that probably only a conservative 10 percent of Y2K problems would ever be known.[9]

In an interesting déjà-vu paragraph, Hyatt writes: "This Y2K-is-a hoax perspective is, at best, a result of extremely shallow research and, at worst, a product of deliberate disinformation. Either way, it is untrue, unhealthy, and dangerous. Those who believe it and do nothing to prepare are putting themselves and their loved ones at tremendous risk."[10]

Hyatt described several known glitches in mid-1999: New Federal Aviation Administration Y2K-compliant software installed at Chicago's O'Hare International Airport was so filled with bugs that air traffic near misses ultimately grounded 100 United Airlines flights; 4 million gallons of sewage was spilled in Van Nuys, California, when a Y2K test closed a key gate; thousands of British passport issuances were delayed due to computer malfunction; Pueblo, Colorado's 911 emergency system Y2K test failure; eight-figure checks were being issued to Monroe County, Indiana, residents; 200,000 New Jersey welfare recipients received $58 million in early food assistance payments; the *Washington Post* was unable to renew subscriptions due to noncompliant software; numerous other issues considered more of an inconvenience than a danger.[11]

An Associated Press article published in the *Boston Globe* mid-January noted that half of the world's total repair costs were spent by the United States: $100 billion. At that time a CBS poll found that two-thirds of Americans believed the expense was worth it, with 98 percent indicating they experienced no Y2K-related problems.[12] Second-guessing was on the short horizon, however.

Good news was also published in the *Insurance Times* in mid-January 2000, reporting that the insurance industry "sails into new millennium with no major problems." Of the $840 billion in total life, health, and casualty/property premiums written, all but a handful of insurers experienced no Y2K software problems.[13] The American Bar Association drew the same conclusion but from the opposite perspective; the uneventful millennium transition "takes the wind out of the sails" of the expected Y2K litigation boom.[14]

The 20/20 hindsight began almost immediately. In a lengthy article in the *New York Times* on January 9, "Smooth Entry of 2000 Is a Puzzle" by veteran Y2K reporter Barnaby Feder, John Koskinen was featured and photographed testifying before Congress in March 1998. The article pointed to the lack of Y2K chaos with demands to have explanations, especially in light of reports of so many countries' understood lack of preparation. I will repeat earlier opinions that corporate management judging overpreparation (and associated costs) is deemed worse by boards of directors than underpreparation and estimation. The *Times* article suggests exactly that.

But it is a reasonable question—why, if other countries were so underprepared, did they escape Y2K seemingly unscathed? What didn't the United States know? Explanations from experts like the Gartner Group implied a general scarcity and vagueness to Y2K readiness data from overseas, as well as suspecting that older, less date-sensitive vulnerable software was in use in many countries thought earlier to be woefully underprepared for the new millennium. The entire article could be summed up in one of the final paragraph's comments: "whether it will be possible to figure out why things went so smoothly."[15]

This should strike everyone as amusing as it strikes me; demands for explanations why things worked out so well! But don't forget that those who were making those demands have but the faintest understanding of the backroom technology efforts to make the software Y2K compliant.

I cannot leave the January *Times* article without one more note: insurers seemed to be on standby to be blamed for it all: "[I]f courts were to decide insurers were liable for the money companies spent to avoid problems, the insurers would undoubtedly cite the success of laggards and low spenders as a sign that budgets for American companies were needlessly bloated."[16] Oh, your honor, I take exception!

The story continues with a February 28, 2000, *Times* article, again by Barnaby Feder, looking at leap year's February 29, 2000. On February 28 there could only be speculation that Y2K fixes had also prepared for the extra day in February. Feder noted that the 1996 leap year had shut down the Belgian Stock Exchange and the Arizona State Lottery, among others. The article cited Canadian Y2K expert Peter de Jager contending that there were over a million Year 2000 malfunctions as of the end of February 2000. The failure of Japan's Shika nuclear plant reporting measurements of radiation levels, building security systems failures including the Canadian government's National Defense Headquarters in Ottawa, and France's military satellite communications failure were noted, along with the Pentagon's spy satellite image processing failure.[17] These were no small issues to point out to the "non-event" believers.

In conclusion, it is my studied opinion that the "windowing" technique of redating files ten years earlier for processing with rebuffering to correct post–01/01/2000 dates for document publication was responsible for the stunning success of the Y2K global crisis. With ten years to analyze, recode, and test software, the millennium deadline pressure was avoided. Subsequent Year 2000 problems occurring in 2010 were corrected in the normal course of software maintenance and routinely not identified as Y2K issues.

That is not in any way to underestimate the monumentally successful labors of America's and the globe's technology personnel to either recode date-sensitive software, develop techniques for more time, or replace older computer systems. A spectacular achievement of talent, management, and cooperation.

AFTERWORD

This book was written to chronicle what I have long called Y2K: the "greatest cooperative effort undertaken by humanity." It has been a journey of reading thousands of documents saved from the 1990s about the solutions, efforts, fears, determination, and grit of the people who labored—labored, but did not know the outcome of their endeavors. What emerged from those efforts had several facets: the usual commercial and economic rivalries; national, state, and local government response to internal and external Y2K threats; both like-minded and disparate approaches to Y2K software solutions; and the universality of Y2K's impact on every facet of business, community, and personal life. As we mark the twenty-fifth anniversary of the millennium turn and the beginning of the twenty-first century, a look back seemed fitting. Indeed, lessons can be learned from the successes of those efforts.

Most of my records review was pro forma and routine. But finding news of the United States sharing early-warning satellite data with both Russia and China leading up to January 1, 2000, while reportedly suffering a Y2K software problem causing a three-day loss of our national reconnaissance data on January 1, GMT, was alarming. It's still astonishing to consider that those two coincident circumstances did not result in a catastrophic consequence.

Ultimately the facts of this story point toward a generally cooperative effort, or perhaps a compatible self-interest, among those entities most responsible for ensuring the nation's and the world's economies held, as well as the basic infrastructures supporting life and labors. And the center did hold! So, what use can be made of the changing complexities of solutions to a single, static, nonchanging date change? What learning curve can help us to develop a roadmap for crises known and as yet unknown?

A ROADMAP FOR CRISIS MANAGEMENT

There are finite but innumerable crisis management advisories for every conceivable situation and condition. Is there anything to the millennium date change that might be distinctive, even unique, to lend insight into why this particular crisis ended so quietly as to be called a "non-event" by most people? I think the answer is yes.

With respect to the Year 2000 computer date change, management in every sector from government, medicine, finance, retail, transportation, and certainly computer systems, understood that 01/01/2000 presented a direct challenge to themselves specifically. I cannot emphasize this enough. This was not just a "What are *they* going to do about it?" scenario, but a "We need to be sure *we* are ready" position. The risk of internal exposure is a great motivator.

A very careful analysis of whom a crisis involves is necessary to engage only those critical resources most affected (i.e., motivated) to find a solution. Monday morning quarterbacks and media pundits should not find their way onto the critical path on any crisis solution. Current times make this much more difficult, I know, as does the public's astounding lack of attention to reliable media reporting. But consider how the handling of the millennium Y2K issue was reported: Y2K responsible parties with official titles confirming their responsibilities, both legal and ethical, were consistently sought and cited.

A team of talented professionals is always front and center for solutions to crises. Y2K was a countercultural example of reverse ageism, where retired COBOL programmers were not only in high demand, but negotiating enviable wages. While the need for the skills of those very senior computer programmers seems obvious, it is wise to pause to consider that innovative solutions to new problems take many forms.

It's not the best-kept secret that regulatory issues have long been considered a real encumbrance on our American innovations and abilities to create and build. However, in the 1990s financial sector, regulators were crucial in maintaining our vital banking system with early Y2K compliance procedures, guidelines, and audits. The same was reported for Wall Street securities regulation. Partnering with regulators, disagreeable as this sounds, is arguably an important part of overall success.

I have also quoted psychologist Ramsay Raymond, who noted that we are a culture held hostage for fear of litigation. That fear, too, was and remains a great motivator; again, a motivator in every sector of commerce and public life.

Stressing the crisis universality of the millennium date change in every sector notwithstanding, the success of Y2K innovation, remediation, and compliance can be unequivocally attributed to the dedication, grit, and hard work of our nation's computer professionals.

We are a culture that is very good at solving problems at the eleventh hour. I hope this book's narrative serves as a benchmark, a roadmap, to approach future crises, knowing that many current challenges, such as climate change and staggering technological advances, are awaiting our considered and timely response. All crises are elements of their time. But with any large-scale challenge at any point in time, cultural and human factors are the drivers of solutions, even more than the influences of technological and scientific progress, in my opinion. The last century leading up to the new millennium had a reliable source of facts from respected media. Facts that were not fabricated for political purposes. Reports of both Y2K hysteria and nonchalance were presented factually, without advocacy to the American public. However, recent references to Y2K in the media have repeated some of the old tropes without pausing to defend or explain. It is time to revisit the facts.

National leaders, professionals, and experts have customarily been in charge of determining solutions to our great challenges. Our very survival depends on that. The people passing judgment on the success or lack of it should be well equipped to assess the experts—those professionals doing the intense hard work to solve our problems. I have described the competing commercial influences that were part of the successful Y2K computer conversion to the Year 2000. I have described our governments' good-faith attempts to assist efforts to meet the millennium without interruption. Those major factors remain today. Politics and power have changed radically.

As former Maine Congressman Tom Andrews stated, "Are we sufficiently covenanted together to do what we so desperately need to do?" And I leave you with that question.

Thank you for reading this book.

ACKNOWLEDGMENTS

Embarking on this project in 2020, I started by drawing the first newspaper clipping from one of the three file boxes containing Year 2000 contemporaneous papers I had saved. Summarizing what had been written quickly became mind-numbing and I abandoned the project as a hopeless approach. When I joined the Reverend Doctor Carl Scovel's Zoom announcement release of his book *I Do* early in the pandemic, he related a story in his inimitable style. His story resonated with me, with vivid memories of Doris Kearns Goodwin so generously speaking to Concord audiences about her work, recounting how she told a story. In particular, Doris relating her account of those people who did not know the outcome of the issue, of the crisis. And there I had it. A story; a story of those who did not know the outcome—in my case, the outcome, the solution to the global Y2K software problem. My reading took a sharp turn toward finding the story; the attitudes, the influences, the greed and generosity—yes, the humanity—of those who did not know the outcome to how the new millennium would enter on January 1, 2000. When I had related my earlier work to acknowledged international intellectual property expert, attorney Lois Wasoff, she said simply, "Write the book." And that stayed with me.

During the next three years chronicling the work of the experts, Capers Jones, a technology and prominent Year 2000 authority, was more than generous with his work and advice. I remain grateful to him for his encouragement and assistance. My long-time business associate, Howard Zaharoff, Esq., whose continuing friendship over the years has meant a great deal to me, helped me with researching the outcome of the Cincinnati Y2K case. While the outcome of that case still lingers, my gratitude to Howard remains. Leading up to my London trip, I asked the Insurance Library of Boston to send me a number of insurance articles related to Year 2000 issues, which were invaluable to me in my work. Finalizing this book in 2023, again the Insurance Library helped to fill in gaps for late 1990s articles that had been lost, for which I especially thank Senior Research Librarian Sara Hart. Looking back to that pre-millennium period, it is important to offer gratitude for the support of both Linda and Michael Graesser—Linda for her steady support in the agency during my absences for Y2K engagements, and Michael for his assistance with strategies and conclusions.

This project would not have been possible without the sage counsel of my agent, Ken Lizotte, of emerson consulting group, inc., as well as emersongroup's talented editor Elena Petricone, whose exceptional perspective amazes me. Of course I include Jake Bonar, my acquisitions editor at Prometheus Books, whose diligence and advice have been invaluable.

While likely belonging at the beginning, family is customarily left to the end to acknowledge and thank. I would not be who I am today, nor the author of anything much, without the support of my amazing partner and husband, Professor Rick Frese. He left me alone to read and to write—an extraordinary man in our culture.

APPENDIX

THE NORTHEAST'S LEADING NEWSPAPER FOR THOSE WHO BUY, SELL AND SERVICE INSURANCE

March 3, 1998
Vol. XVII No. 5

Insurance & Benefits Technology Vendor Profiles
Page 13

Employment Opportunities
Pages 21-23

28 PAGES

IN THIS ISSUE

Do's and don'ts of automation

Agencies today have no choice: they must upgrade their automation systems. InsuranceTimes asked the experts: Where to start? What to expect? How much will it cost?

by Penny Williams
Special to InsuranceTimes

Technology is changing so rapidly that by the time an agency goes through the process and installs an automation system today, the system is already obsolete. Experts note that agency automation systems become old quickly, within as little as 18 months.

If that reality were not enough, the Year 2000 challenge is forcing many agencies to upgrade and change. Experts warn that only a matter of months remain for agencies, companies, vendors, suppliers — everyone and everything that is date-sensitive — to become Year 2000 compliant. Agencies not Year 2000 compliant, experts say, will see their businesses screech to a halt and most will not survive.

Agencies needing to upgrade today confront a consolidated vendor market that has left only two major vendors — AMS and Applied Systems — along with a handful of smaller, less well known vendors.

InsuranceTimes tapped the brains and experience of some leading consultants and agents to

continued on page 26

Millennium liability slated to hit Insurers, agents, lawyers and more

by Penny Williams
Special to InsuranceTimes

In addition to investing millions of dollars to update its own technology, the insurance industry must gird for an explosion of lawsuits surrounding the failure of computer systems to be Year 2000 compatible.

Also, the directors and officers of high tech firms as well as the attorneys who drafted technology contracts should arm themselves, as they will be among the potential targets of Year 2000 lawsuits.

"The liability is limited only by the limits of one's imagination," maintains Nancy P. James, owner of N.P. James Insurance Agency in Concord, Mass., who recently addressed the American

continued on page 26

Captive agents seek contract protections

Independent agents may think they have it tough — but captive agents have their problems, too.

by Chris Mahoney
Special to InsuranceTimes

There are times when David Donaldson doubts if Nationwide really is on his side.

Donaldson has been an exclusive representative for the Bloomington, Ill.-based insurance company since 1989, when he took over his father-in-law's agency in Manchester, Conn.

"Was my father-in-law happy with Nationwide? Yes. But was it a different company from what it was five years ago? Yes," Donaldson said.

"Will it change? I hope so."

Donaldson accuses the insurer of a number or questionable practices — and he's not alone. The insurance giant is the source of war stories from a number of its Connecticut-based captive agents.

They dredge up the company's infamous 1995 minimum standards program, and the termination threats they said were leveled at them if they didn't meet company-set quotas. They claim that such measures force some agents to resort to churning, tie-ins and other illegal acts to meet sales goals. Nationwide dropped its minimum standards program in 1997.

They cite Nationwide's legal woes in other states. In Florida, for example, the insurer withstood allegations that it pressured its agents to exclude certain customers based on age, sex, marital status or occupation. Nationwide denied doing anything of the kind, but agreed to pay the state a $100,000 settlement.

They question the company's widely touted five-year development campaign, which includes plans to use "multiple distribution channels" to attract new customers in hopes of doubling the company's revenues by 2002.

And some have individual stories: the company is accused of financing an agency which now competes

continued on page 25

NH bill to deregulate large commercial risks advances

by Penny Williams
Special to InsuranceTimes

CONCORD, N.H. — A bill to deregulate large commercial insurance risks in New Hampshire has passed the state Senate and is now in the House.

Industry observers believe that enactment of the legislation could occur soon, although the House may make some changes to the Senate version.

Independent agents and insurers support the measure, which does not affect commercial auto or workers compensation lines.

"There is general support for the bill and I know of no opposition to the general concept," com-

continued on page 17

Travelers' entry turning point in Internet insurance sales

NEW YORK (AP) — Travelers Property Casualty has become the first major insurance company to allow U.S. consumers to purchase automobile insurance directly over the Internet, initially offering the service in Alabama.

So far, insurance buyers have been able to use the Internet mostly to obtain quotes and the names of area insurance agents from whom they could purchase a policy.

Travelers' new product, available immediately to Alabama customers on Intuit Inc.'s Quicken InsureMarket Web site, will allow them to complete the transaction over the Internet and obtain coverage beginning at midnight on the following business day. To close the

continued on page 17

Nancy P. James is quoted on the front page of the *Insurance Times*, March 3, 1998.
Courtesy of the Insurance Library of Boston.

NEWSLETTER n.p. james insurance agency

Volume 1, Number 2
May 1997

Nancy James Addressing American Bar Association, August 1

The American Bar Association has invited Nancy James to speak at its Annual Convention August 1, 1997, in San Francisco, CA. Ms. James will sit on the Technology Panel discussing Year 2000 liability issues.

Year 2000-Your Liabilities for the World's Greatest Deadline

Year 2000 is one of the hottest issues in both technology and business, as every enterprise scrambles to become Year 2000 compliant. The consequences of non-compliance can not only result in internal chaos but angry and litigious clients!

Often described in apocalyptic terms, carrying a global price tag of $600-$700 billion, Year 2000 solutions cannot be delayed, rescheduled, or ignored. Sales guarantees, service contracts, implied and contracted warranties, need to be carefully examined for client assumptions on all products and services having Year 2000 compliance issues.

Attorneys, accountants, and insurance professionals are becoming conscious of advisory matters to our clients. From the insurance side, only Technology Errors & Omissions coverage will respond to those consequential damages (financial injury) created by Year 2000 problems.

Now is not too soon to discuss your errors and omissions issues with a technology insurance specialist.

Technology Errors & Omissions Coverage

A new product, underwritten by USF&G, offers technology Design Errors & Omissions Liability Insurance with annual premium rates starting at $2,500. Call us for details.

Announcement

The N.P. James Insurance Agency is pleased to announce that it is now one of the four designated agents for the ***Massachusetts Software Council's*** sponsored insurance program through USF&G.

Cyberspace Risks

In the year 1450, 10,000 hand copied texts existed in all of Europe. By 1500, 9,000,000 books were in print. That was the first great proliferation of knowledge!

Cyberspace liability is not fundamentally a technology issue. It might as well be a monk on a stool with a quill pen copying biblical texts. Print and, later, broadcast media have historically faced the same liabilities as cyberspace. To whit:

- copyright infringement
- defamation
- disparagement of an individual's reputation
- infliction of emotional distress
- infringement of title, name, or mark
- invasion or infringement of the right of privacy
- patent infringement
- piracy
- plagiarism
- product disparagement
- unfair competition

The staggering difference in cyberspace is that territorial boundaries are gone. Countries which have customarily prohibited specific materials from publication can no longer prevent entry. Second and third world countries are struggling with the necessity of compliance with industrialized copyright, trademark, and patent laws, while our common domestic websites circle the globe, indifferent to national jurisdictions.

The risk management of corporate liability exposures must have top priority.

We can help.

A Publication of
N.P. James Insurance Agency • 33 Bedford Street • Concord, MA 01742 • (508) 369-2771 • FAX (508) 369-2778 • © 1997
Celebrating 15 years in business

Courtesy of the author.

PRESIDENT'S COUNCIL ON YEAR 2000 CONVERSION

CHAIR

September 17, 1998

Ms. Nancy P. James
N.P. James Insurance Agency
33 Bedford Street
Concord, MA 01742

Dear Ms. James:

Thank you for your letter of September 11 and your suggestion that we establish a "Millennium Mediation Council." One of the 34 working groups the Council has established is beginning a dialogue with insurance industry umbrella groups and I look forward to discussing with them various aspects of the questions of allocating costs relating to year 2000 events.

Best wishes.

Sincerely,

John A. Koskinen
Assistant to the President

ROOM 115 • OEOB • WASHINGTON, D.C. 20502 • (202)456-7171 • FAX(202)456-7172

President Clinton's "Y2K Czar," John Koskinen, telephoned and replied to Nancy P. James regarding her recommendation of a "Millenium Mediation Council." *Courtesy of the author.*

As seen in

THE JOURNAL OF NEW ENGLAND TECHNOLOGY

MASS High Tech

THE JOURNAL OF NEW ENGLAND TECHNOLOGY

www.boston.com/mht

JULY 13 - 19, 1998

VOLUME 16, ISSUE 28

'War game' aims to prevent legal battles for insurer

BY NANCY P. JAMES

Nancy James

Shark-infested cyberspace ... at least it should seem so for any Internet service provider, carrier, or anyone else with a website. Cyberspace court cases have been making the news for several years now, and indications to date suggest that an absolute, or close to absolute, liability can be imposed! It appears that not only U.S. statutes but also international laws, including language requirements, have to be met.

This, of course, raises interesting questions for the insurance industry, which must judge such risks with no actuarial history of risk exposure or losses.

Mainspring Communications, Inc. of Cambridge is an advisory firm that focuses on accelerating Internet business through rapid strategy alignment and strategic execution and evaluation. In early 1997, seeing Internet risks everywhere, Mark Verdi, Mainspring's vice president of finance, contacted me, as a technology risk specialist and principal of N.P. James Insurance Agency in Concord, to put together a program of protection as Mainspring embarked on a brand new Internet service concept.

Working with Verdi and an underwriter from a media and publishing risk insurer — not, interestingly, a technology insurer — we structured the first liability-based cyberspace "war game" for Mainspring. We believe that this was one of the first times a software company allowed a potential insurance underwriter to test their website to determine what liability and risk vulnerability they were assuming as a Web provider.

Our objective was to determine what kind of coverage was necessary to protect Mainspring as they moved forward with their Internet strategy. Uneasy that the test might actually accomplish some of these unwanted results such as copyright infringement, the insurer's underwriter claims counsel requested and received elaborate hold-harmless agreements against such eventualities.

Verdi arranged for beta entry codes for both the underwriter and me to see how secure the site was from a loss risk standpoint. The plan for the war game was to see if copyright-protected material could be downloaded by the site visitor, if offensive material could be posted on the chat board, and what other areas might lead to possible misuse, offense, and/or litigation.

Claims counsel in the Midwest, and Verdi and I in Cambridge, were opened to the site and connected by speaker conference telephone. We centered our discussion on the areas in which the insurance underwriter and I were looking, as Verdi explained the content and site intent for on-line users.

Mainspring, as is common among many sophisticated website and Internet services, had a Games and Contests section. When visiting that site section, claims counsel asked whether Mainspring had sought approval from all 50 state gaming commissioners. Prudence dictated that the section be eliminated for the initial live-entry date because of that issue.

This same huge dilemma exists for all types of licensed parties. Professionals such as doctors, lawyers, insurance agents and brokers, accountants, and many others are licensed on a state-by-state basis. The consequences of advertising and soliciting in locales where licenses are not held looms darkly on the horizon, especially if an interactive site encourages direct response from global market consumers.

Following that thread leads to jurisdictional questions. Currently, it appears that jurisdiction, unless otherwise noted, is where the transaction was initiated or where the interactive site was first activated. So we must ask if doing business globally means we must be ready to defend suits in jurisdictions a hemisphere away.

As a result of our "war game," Verdi, with my assistance, judged what content was more exposed to litigation and what content to drop, as well as the best program of risk transfer to the insurer. Mainspring, wisely looking very hard at the liabilities they might be assuming as a cutting-edge, Web-based service provider, opted to purchase insurance coverage against many of the intellectual property perils other publishers face: copyright infringement; defamation; disparagement of an individual's reputation; infliction of emotional distress; infringement of title, name, or mark; invasion or infringement of the right of privacy; privacy; plagiarism; product disparagement; and unfair competition.

Patent infringement is an area considerably more complex and costly to underwrite and insure, and was not within the scope of Mainspring's coverage needs. Obscenity, a real problem as the Web reaches across borders into conservative political areas, is considered criminal in all societies and is simply against public policy to insure. Therefore, anyone with an interactive Web site should be aware that defending against postings of an obscene manner will be defending without their insurer.

Happily, in this cyberspace war game, all sides came out winners.

Nancy P. James is the president of N.P. James Insurance Agency in Concord, MA where she specializes in risk analysis and insurance for technology-based clients. E-mail her at npjames@compuserve.com.

Nancy P. James describes how the first internet liability insurance policy was produced and issued. *Courtesy of the author.*

N.P. JAMES INSURANCE AGENCY
33 BEDFORD STREET
CONCORD, MA 01742
TELEPHONE (978) 369-2771 FAX (978)369-2778
NPJAMES@COMPUSERVE.COM WWW.NPJAMES.COM

COMMENTARY

THE YEAR 2000 PROBLEM AND INSURANCE COVERAGE

The mere magnitude of Year 2000 problems creates a staggering obstacle when contemplating recovery of Year 2000 computer related failures through insurance claims. The total US insurance industry reserves are reported to be $280 billion. With the worldwide costs of Year 2000 fixes estimated at $600 billion and litigation costs climbing to estimates of $1.5 trillion, a myriad of damage and liability claims simply cannot be expected to be turned over to insurers for defense and reimbursement.

It must be noted that AIG, the only insurer with active Year 2000 coverage policies in the U.S., has closed the market from this date forward due to adverse selection.

This, singularly, like no other issue in history, is not only an opportunity but a necessity for professionals to work together to direct every available asset toward solving the Year 2000 conversion. Courts must cooperate, the bar must direct its efforts toward bringing its clients into compliance without litigation, congress and legislatures must redirect from municipal immunity to public assistance, while the public and private sectors join hands to solve the problem, short and long term.

This undated mid-1990s commentary by Nancy P. James pleads for cooperation rather than litigation with regard to Year 2000 computer problems. *Courtesy of the author.*

SEPTEMBER 15, 1998 InsuranceTimes 7

GUEST EDITORIAL

A Modest Year 2000 Proposal: *A Millennium Mediation Council*

A PROPOSAL TO AVOID THE COSTS OF LITIGATION SURROUNDING THE YEAR 2000 PROBLEM

by Nancy P. James

Having just returned from London, where the E.U. insurance community asked for a strong U.S. lead on Year 2000 coverage concerns, I am offering a modest proposal which breaks all our traditional business paradigms.

Just a year ago, on August 1, 1997, I addressed the American Bar Association Annual Meeting on the Year 2000 subject, concluding my early predictions on Year 2000 influences by asking the bar to join, as they have never before done, with their insurance and accounting professional colleagues in solving the Year 2000 problem for our clients.

Today it is a categorical imperative! Current estimates of this $600 billion global problem will exhaust twice over the U.S.'s estimated $280 billion domestic insurance reserves without even touching the $1 - $1.5 trillion the bar expects in litigation costs and fees.

The E.U. community, without equivocation, concurs that U.S. courts will likely deny our Year 2000 exclusions. Exclusions denied assume coverage on the basic, unendorsed policy. Letters of clarification from carriers go a long way toward client understanding of coverage, but ultimately, no affordable independent commercial Year 2000 coverage exists in the states

today. And, coverage areas have yet to be tested in the courts, which, in my judgment, will rival patent litigation costs of $500,000 pre-trial and $500,000 at trial.

We still have an opportunity to solve this situation before Year 2000 arrives, and allocate every precious available resource toward the solution and away from adversarial dispute expenses.

A national body. The Year 2000 Mediation Council, including insurance, accounting and law professionals, must settle coverage questions and limits ahead of time; mediation before the fact for this unfortuitous event. Claims limits within categories need to be resolved, as do uncovered areas of Year 2000 loss; attorney's fees; and stated, common defended areas (fire, bodily injury). These areas can be quantified; we can advise our clients precisely; and we can be a valuable part of a global solution to Year 2000. Thus, we will be able to direct U.S. resources toward technical Year 2000 solutions and fail-safes.

We can lead the E.U. and global insurance communities in a rational strategy.

And, we must urge our industry and government officials to support us in our efforts. ◻

James is president of N.P. James Insurance Agency in Concord, Mass. Telephone: (978) 369-2771 Fax: (978) 369-2778
NPJAMES@COMPUSERVE.COM
WWW.NPJAMES.COM

Nancy P. James offers a guest editorial to the *Insurance Times* promoting a Millennium Mediation Council for Year 2000 technical solutions. *Courtesy of the author.*

› Files from NPJames Ins. computer › Users › Nancy › Documents › FILESYS › MONTH 1999

Name	Date modified	Type	Size
01JAN	9/27/2019 7:15 PM	File folder	
02FEB	9/27/2019 7:15 PM	File folder	
03MAR	9/27/2019 7:15 PM	File folder	
04APRIL	9/27/2019 7:15 PM	File folder	
05MAY	9/27/2019 7:15 PM	File folder	
06JUNE	9/27/2019 7:15 PM	File folder	
07JULY	9/27/2019 7:15 PM	File folder	
08AUG	9/27/2019 7:15 PM	File folder	
09SEPT	9/27/2019 7:15 PM	File folder	
10OCT	9/27/2019 7:15 PM	File folder	
11NOV	9/27/2019 7:15 PM	File folder	
12DEC	9/27/2019 7:15 PM	File folder	
-98HOME.REN	7/13/1989 7:37 AM	REN File	15 KB
-BAILEY.BOR	7/20/1989 3:58 PM	BOR File	10 KB
-CHUBB.LST	12/1/1989 4:06 PM	LST File	10 KB
-COM-REV.LST	7/1/1989 7:24 AM	LST File	13 KB
-COST.SUM	11/1/1989 4:02 PM	SUM File	9 KB
-CYB-Y2K-LRT	12/2/1998 12:33 PM	File	12 KB
-D&O-DOS.IER	7/1/1989 1:59 PM	IER File	14 KB
-escrow-bank-let.DOC	1/10/2000 11:21 AM	Microsoft Word 9...	11 KB
-EXPAND.SUM	1/31/2000 9:39 AM	SUM File	45 KB
-FLOOD.REN	6/9/1989 7:18 AM	REN File	13 KB
-GLOBAL.SGN	2/23/1989 4:44 PM	SGN File	10 KB
-HO&UMB.REN	5/6/1989 2:41 PM	REN File	15 KB
-L-HEAD.DOC	1/6/1999 9:25 AM	Microsoft Word 9...	9 KB
-LITTLE.HED	4/1/1989 11:03 AM	HED File	10 KB
-RCC-RCD.FLD	12/16/1998 12:10 PM	FLD File	31 KB
SAWYER.RMV	6/21/1989 2:13 PM	RMV File	10 KB
-Y2K-REMIND	10/25/1989 1:22 PM	File	9 KB

Nancy P. James discovered these pictured 1989 dates in her computer files after embarking on writing this book. Yet these 1989 dates are impossible since James did not have Microsoft Windows until 1994. Despite outreach, including to Microsoft, these 1989 dates still have no explanation. Author notes that the actual file documents were all dated 1999.
Courtesy of the author.

NOTES

INTRODUCTION: "Y2K"

1. K. C. Bourne, *CGL Reporter,* Fall 1998, citing *Year 2000 Solutions for Dummies* (IDG Books, 1997). Note that computer memory cost reports varied widely. A February 1998 *Money* magazine article by Andrea Rock and Tripp Reynolds ("The Year 2000 Bug," 50) cites mid-1970 megabyte costs at $600,000 with early 1998 cost at 10 cents.

2. John Diamond, "Satellite Post Fixed after Y2K Glitch," *Boston Globe*, January 13, 2000; "Y2K Bug Bit Pentagon Satellites at Key Time," *Boston Globe,* January 4, 2000.

3. Fred Kaplan, "Military on Year 2000 Alert: 2000 a Computer Time Bomb for Military," *Boston Globe,* June 21, 1998.

4. Nancy James and Barbara O'Donnell, "A Primer on the Year 2,000 [*sic*] Bug, A Role for Careful Counsel," *Boston Bar Journal,* May/June 1997. (In early articles referencing Year 2000, a comma was often mistakenly inserted: 2,000. This *Boston Bar Journal* piece was one of the first articles published in a law journal.)

5. M. A. Nelen, "The Y2K Problem: How to Survive Meltdown AND the Lawsuits," *Mass High Tech*, July 13–19, 1998, 9.

6. American Bar Association, Committee on Insurance Coverage Litigation, March/April, 1998, Coverage publication, "The Y2K Time Bomb—Policyholders Run for Coverage" by Terry Budd, Esq., and Curtis B. Krasik, Esq., Kirkpatrick & Lockhart, LLP, Pittsburgh, 3–17.

7. *Chronicle-Herald*, Halifax, Nova Scotia, July 12, 1996.

8. Anne Colden, “Insurers Differ Over Paying Y2K Claims,” Associated Press, May 27, 1998.

9. Douglas G. Houser and Linda M. Bolduan, Bullivant Houser Bailey (Portland, OR), “Across Frontiers: Practice and Procedure Y2K: Assisting Clients with Global Vendors,” *Defense Research Institute*, Winter 1999, 7, quoting Chris Allbritton, “More Trillion-Dollar Bugs Await,” Associated Press, October 12, 1998.

10. “Y2K to Be Top Issue in Directors and Officers Liability,” *Best's Review* via NewsEdge Corporation, December 7, 1998.

11. Andrew Q. Barrett and Mark B. Fenton, “Beyond the Millennium Bug: What Investors Can Expect Now,” Salomon Smith Barney, June 1999.

12. Trevor Thomas, “First Y2K Suits May Start Flood,” *National Underwriter*, January 4, 1999, 3, 14.

13. “Please Panic Early,” *Economist*, October 4, 1997, sent to me by Dudley Cunningham, Boston Private Bank.

14. Suzanne Sclafane, “Best Expects Big Fall in Insurer Total, ROE,” *National Underwriter*, January 4, 1999, 29.

15. Andrew M. Pegalis, with Professional Risk Managers Services (Arlington, VA), “For Risk Managers, the Year 2000 Is Now,” *Business Insurance*, December 23/30, 1996, forwarded from the Insurance Library, Boston. Pegalis was quoting U.S. Senator Daniel P. Moynihan on the global Y2K cost estimate.

16. Bruce D. Berkowitz, “‘Y2K’ Is Scarier Than the Alarmists Think,” *Wall Street Journal*, June 18, 1998; thanks to Penny Williams of the *Insurance Times*.

17. Anne Colden, “Year 2000 for Insurers: Armageddon or No Big Deal?” *Dow Jones News Service* via Dow Jones, May 11, 1998.

18. Nancy P. James, “‘War Game’ Aims to Prevent Legal Battles for Insurer,” *Mass High Tech*, July 13–19, 1998. The article is reproduced in the appendix.

19. Robert Duggan, Stanley Keller, Ann C. King, and George Ticknor, “Client Alert: ‘Year 2000’ Issues Require Attention by Lenders and Investors,” *Palmer & Dodge, LLP, Newsletter*, November 1997.

20. Katheryn M. Welling, “Millennium Bug: Will Year 2000 Fixes Create Giant Sucking Sound?” (interview with Fred Hickey, publisher of *High-Tech Strategist*), *Barron's*, July 14, 1997, 29.

21. Edmund X. DeJesus, “Year 2000 Survival Guide,” *Byte Magazine*, July 1998, 52–62.

22. Barnaby J. Feder, “The Town Crier for the Year 2000,” *New York Times*, October 11, 1998.

23. Alison Harris, “‘Year 2000’ Can Offer Silver Lining: Forces Technical and Business Groups Together,” *Service News* reprint, the Newspaper for Computer Service & Support, June 1996.

24. Andrew Pollack, “For Coders, a Code of Conduct: 2000 Problem Tests Professionalism of Programmers,” *New York Times*, May 3, 1999, C1, C12. Pollack interviewed Paul A. Strassman, former chief information officer, Defense Department and Xerox Corporation; and Leon A. Kappelman, associate professor of

business computer information systems, University of North Texas, cochair of the Society for Information Management's Year 2000 working group.

25. David Clay Johnston, "Taxes: I.R.S. Anticipates Year 2000 Well Ahead, Early in 1999," *New York Times*, January 3, 1999.

26. Leland G. Freeman, "Leadership: Will the Next FDR Come Forward," *Software Magazine*, October 1998 ("This Is a Test" issue, the last of the Year 2000 issues). Freeman was vice president of the Source Recovery Company, LLC, specializing in retrieval of lost computer code (lfreeman@mci2000.com).

27. Peter Worrall, editor, "Year 2000—Just a Toss of the Dice?" *Insurance Specialist* (an independent publication in the United Kingdom), in ACORD, November 4, 1997, 15–16.

28. Capers Jones (chairman, Software Research, Inc., Burlington, MA), "Abstract: Dangerous Dates for Software Applications," March 9, 1998, 13 (sent to Nancy James on March 14, 2022, with permission to use).

29. Jones, "Abstract: Dangerous Dates."

30. M. J. Anderson, "On the Increasing Danger of Cyberattacks: Generally Recognized as the First Known Cyberattack by One Nation Against Another" (Russian hackers against Estonia in 2007), *Boston Globe*, April 24, 2022, N10–N11.

31. Anne E. Kornblut, "Senate Y2K Watchers Sound Muted Alarm," *Boston Globe*, March 3, 1999, A3.

32. Berkowitz, "'Y2K' Is Scarier Than the Alarmists Think."

33. Barnaby J. Feder, "Fear of the Year 2000 Bug Is a Problem, Too: Year 2000 Bug Meets People Problem: Surprising Early Outbreak of Panic," *New York Times*, February 9, 1999.

34. "Time Is Running Out on Passage of Y2K Liability Bill," *National Underwriter*, June 28, 1999 (editorial comment citing Federal Reserve Board Chairman Alan Greenspan's testimony of early 1999 to the Senate hearings on Y2K conducted by Senator Robert Bennett, R-Utah).

35. Matthew Schwartz, "Supply Chain Nightmare," *Software Magazine*, "This Is a Test" issue, International Suppliers, October 1998 (the last of the Year 2000 issues).

36. William Ulrich, "Q&A: Ask Dr. 2000," *Software Magazine*, October 1998, ("This Is a Test" issue), tsginc@cruzio.com.

37. C. Ramsay Raymond, owner, The Dreamwheel (Concord, MA), letter of July 13, 1999, to Nancy P. James, owner, N. P. James Insurance Agency (Concord, MA).

38. Capers Jones, Abstract: "Probabilities of Year 2000 Damages—Version 3," February 6, 1999; and Abstract: "Probabilities of Year 2000 Damages," February 27, 1998.

39. Gary North's Y2K Links and Forums, "FEMA Holds Closed Conferences. Were You Invited?" *Mirror*, February 19, 1999, www.y2k-links.com/garynorth/3903. Via Valerie Kinkade, Concord Neighborhood Network.

40. *Mealey's Litigation Report*, April 1998.

41. *Sunday New York Times*, CNBC advertisement, January 2, 2000, 11.

CHAPTER 1 READY OR NOT

1. Barnaby J. Feder, "Clinton Optimistic About Year 2000: After Report on Readiness, Some Call His Predictions Too Rosy," *New York Times*, November 11, 1999.
2. Robert Sam Anson, "12.31.2000: The Y2K Nightmare," *Vanity Fair*, January 1999.
3. Anson, "12.31.2000."
4. Anson, "12.31.2000."
5. Anson, "12.31.2000."
6. Anson, "12.31.2000."
7. "Information Technology 01-01-00," *Government Executive,* mid-1996.
8. Steven Levy, "Random Access," *Newsweek,* June 22, 1996.
9. C. Ramsay Raymond, owner, The Dreamwheel (Concord, MA), letter of July 13, 1999, to Nancy P. James, owner, N. P. James Insurance Agency (Concord, MA).
10. Raymond, letter of July 13, 1999, to Nancy P. James. This passage is Raymond's work verbatim; the punctuation is presented just as it appeared in the source text.
11. Raymond, letter of July 13, 1999, to Nancy P. James.

CHAPTER 2 TECHNICAL SOFTWARE SOLUTIONS TO Y2K

1. Barnaby J. Feder, "Media & Technology: Companies Prepare for Year 2000," *New York Times,* January 4, 1999, C17.
2. "Users, Insurers Grapple over Y2K," *InfoWorld* via NewsEdge Corporation, August 18, 1998.
3. Capers Jones, Chairman, Software Research, Inc., "Abstract: Dangerous Dates for Software Applications," March 9, 1998, 13 (sent to Nancy James on March 14, 2022, with permission to use).
4. Peter de Jager, "Y2K: So Many Bugs . . . So Little Time," *Scientific American*, January 1999, 88–93.
5. Barnaby J. Feder, "Companies Lag on Year 2000 Repairs, Study Says," *New York Times*, May 17, 1999.
6. C. Carl Dodson, CPCU, vice president of client services, CAN Re (Chicago), "A Practical Approach to the Year 2000 Problem," *CPCU Journal*, Summer 1998, 74–76, 78, 79, 81.
7. Dodson, "A Practical Approach to the Year 2000 Problem."
8. Jones, "Abstract: Dangerous Dates for Software Applications," 13.
9. Barnaby J. Feder, with reference to Wattenburg's website www.drbill .org and IBM's website www.y2k0k-solution.com (neither of which are currently

functioning), in "Fixing Year 2000 Computer Problems May Be as Simple as Counting to 16," *New York Times*, November 15, 1999.

10. John Kerr, "Only a Test?" Editor's letter, *Software Magazine*, October 1998 ("Year 2000 Survival Guide" issue).

11. "Year 2000 Compliance Management," Keane product brochure CS-YR2000-1196, Keane's Resolve 2000 Solution, Keane, Inc. (Boston, MA).

12. "The Year 2000: Addressing the Year 2000 Challenge," SAPTM Banking product flyer 50-018-303, 1997, copyright 1997.

13. "PKF Alert," and "RoadMap 2000," PKF Technologies, Inc., member, Pannell Kerr Forster International, Certified Public Accountants & Management Consultants (Honolulu, HI), signed Garrett Kojima, executive vice president and director of marketing (received September 21, 1998).

14. "Year-2000 Solutions Digital Equipment Corp.: AD&I Practice Year 2000 Briefing," Digital Equipment Corporation, December 10, 1996, prepared by D. Harrington & the Year 2000 Team.

15. Withers, "2000," service portfolio, Withers (London), ca. 1998.

16. "Year 2000 Solution Providers Directory and Directory Addendum," Information Technology Association of America, Spring 1997, and May 1997. "ITAA*2000 Certification Program Information Kit," opened October 1, 1996.

17. Pam Derringer, "Year 2000 Watchdog Group Has Formed," *Mass High Tech*, December 9–15, 1996.

CHAPTER 3 CORPORATE PREPAREDNESS, BANKS, AND INVESTING: A BRIEF LOOK AT HOW CORPORATE AMERICA PREPARED

1. Steven Levy, "The 1,000 Year Glitch," *Newsweek*, June 22, 1996.

2. "Please Panic Early," *Economist*, October 4, 1997, sent to me by Dudley Cunningham, Boston Private Bank.

3. David Reich-Hale, "'War Rooms' Get Ready to Combat Y2K Woes," *National Underwriter*, January 4, 1999, 3, 14–15.

4. Ross Kerber, with Thomas C. Palmer Jr., Alex Pham, Ronald Rosenberg, Diane Lewis, and Lynnley Browning contributing, "Racing the Clock," *Boston Sunday Globe*, January 31, 1999, 1, 26–27. This included local compliance graphs for utilities, telecommunications, banks/financial services/exchanges, hospitals/HMOs/insurers, transportation, state/local government, law enforcement, food/retail, and major employers.

5. "High-Tech Firms Book Y2K Rooms," *High-Tech Quarterly*, November 12–18, 1999.

6. Richard Saltus, "Health Systems Near Set for Y2K: Contingency Plans Being Fine-Tuned." *Boston Sunday Globe*, August 15, 1999, 1; "Hospitals Speed Up Y2K Plans," *Boston Sunday Globe*, August 15, 1999, B5.

7. As reported to Nancy James, April 16, 2023, by Thomas Cafarella, master software engineer, Hewlett Packard/Compaq/Digital Equipment Corporation, member of DEC's rapid response team for Year 2000 crisis management.

8. Roger Lowenstein, "Intrinsic Value," *Wall Street Journal,* July 25, 1996.

9. "First Call Hit by Year 2000 Bug: Can't Plug in Earnings Estimate Data for Beyond Start of New Century," *Boston Globe*, Bloomberg News report; date noted only as "1/98," quoting analyst Peter Aseritis of Credit Suisse First Boston.

10. Helen Graves, "Software Companies Strong Growth Spikes with Y2K" (interview with Anne Brennan, president and CEO of the value-added accounting software reseller Klear Solutions), *Women's Business*, August 1999, 11, 25.

11. Alex Sheshunoff, president of the bank consulting and executive development firm Alex Sheshunoff Management Services (Austin, TX), "The Year 2000 Crisis," *Bank Director*, First Quarter, 1997.

12. Michael P. Norton, "State Officials Looking at Ways to Make Banks Think Y2K Is for Real," *Mass High Tech*, December 28, 1998–January 3, 1999, 5.

13. Lynnley Browning,, "Keeping the Faith: Besides Taking Y2K Precautions, Bankers Wrestle with Worries of Consumer Panic," *Boston Globe,* undated early 1999 article.

14. Browning, "Keeping the Faith"; Jonathan Fuerbringer, "Seeing Signals from Investors: Treasury Bill Auction Shows Concern on Risk; Money Managers Reading Investor Signals on Year 2000," *New York Times*, July 27, 1999, Market Place section, 1.

15. "Congress Revs Up to Tackle Y2K Liability Legislation," citing remarks by Karen Shaw Petrou, president of consulting firm ISD/Shaw, NewsEdge Corporation, via Penny Williams at the *Insurance Times*, February 20, 1999.

16. "Taking Stock of Y2K as Deadline Approaches: Federal Reserve's Cathy Minehan Foresees No Reason to Panic," *Women's Business*, August 1999, 6.

17. "The Year 2000 Date Change: What the Year 2000 Date Change Means to You and Your Insured Financial Institution," undated FDIC brochure, Bankers Systems, Inc. (St. Cloud, MN), FDIC-Y2KBRO.

18. "Safety and Soundness Guidelines Concerning the Year 2000 Business Risk," Federal Financial Institutions Examination Council (FFIEC), to the board of directors and chief executive officers of all federally supervised financial institutions, providers of data services, senior management of each FFIEC agency, and all examining personnel, December 17, 1997.

19. "A NOTE ABOUT Y2K," undated Citibank enclosure notice.

20. Untitled, *Business Insurance* via NewsEdge Corporation, February 12, 1999, via Penny Williams at the *Insurance Times*, February 12, 1999.

21. Associated Press report, *Boston Globe*, January 4, 2000.

22. Ara Trembly, *National Underwriter*, October 25, 1999.

23. Rauer L. Meyer, and Steven L. Hock, Thelen, Marrin, Johnson & Bridges, LLP, "Legal Issues and Risk of the Year 2000 Problem," Year 2000 Information Center—Year 2000, June 20, 1998.

24. "Y2K to Be Top Issue in Directors and Officers Liability," *Best's Review* via NewsEdge Corporation, December 7, 1998.

25. Andrew M. Pegalis, Professional Risk Management Services (Arlington, VA), "Year 2000 for Risk Managers, the Year 2000 Is Now," © 1996, reprinted from *Business Insurance*, December 23–30, 1996, forwarded from the Insurance Library (Boston, MA).

26. Pegalis, "Year 2000 for Risk Managers, the Year 2000 Is Now."

27. "Y2K and Your Insurance: The Issues," J&H Marsh & McLennan, mid-1998, 11.

28. Kimberly A. Strassel, "Millennium Problems? Dial Up Your Lawyer: Turn-of-the-Century Computer Fears Spur a Specialty," *Wall Street Journal Europe*, May 16, 1997.

29. Patricia Vowinkel, "Insurers Seen Hit by Millennium Bug Losses," ZDNet, Ziff-Davis, Inc., June 4, 1998. The article quoted Jay Cohen, analyst at Merrill Lynch: "[I]f just 10 percent of the $600 billion fell into the laps of insurers, it would eat up about 20 percent of the U.S. insurance industry's $310 billion in policyholder surplus."

30. "Analysts Seek More Disclosure of Y2K Liability" (reporting on Ernst & Young Director David Holman's address to the Insurance Accounting & Systems Association), *Insurance Accountant*, June 22, 1998.

31. Joan Hartnett-Barry, "CAT CODE Y2K, Computer Date Error Threatens to Derail Business, Cause Claims," *CPCU Journal*, January 1998, 36–38, 79–80. "CAT" is an insurance shorthand term meaning "catastrophe," which is a specific designation for measuring and assigning claims costs.

32. Andrew Q. Barrett and Mark B. Fenton, "Beyond the Millennium Bug: What Investors Can Expect Now," Salomon Smith Barney, June 1999.

33. Barnaby J. Feder, "The Dominant Position of the Gartner Group," *New York Times*, July 5, 1999, C1–C2.

34. Barnaby J. Feder, "Lull in the Party for Year-2000 Stocks: '97 Was a Very Good Year, but Prices Have Run Down with the Clock" (citing Bloomberg Financial Markets Year 2000 Index), *New York Times*, October 25, 1998, 8.

35. Feder, "Lull in the Party for Year-2000 Stocks."

36. Barnaby J. Feder, "As the Clock Nears Midnight: Companies That Dance in the Year 2000 Market Need to Find New Steps Before the Ball Ends," *New York Times*, July 15, 1999, 1, 23.

37. Feder, "As the Clock Nears Midnight."

CHAPTER 4 YEAR 2000 IS NOT COVERED: Y2K AND INSURANCE

1. Ara C. Trembly, "Barristers Battle Over Y2K Liability," *National Underwriter*, July 26, 1999, 2, 38.

2. Penny Williams, "Millennium Liability Slated to Hit Insurers, Agents, Lawyers and More," (front page article interviewing Nancy James), *Insurance Times*, March 3, 1998.

3. Williams, "Millennium Liability Slated to Hit Insurers, Agents, Lawyers and More."

4. Williams, "Millennium Liability Slated to Hit Insurers, Agents, Lawyers and More."

5. "Insurers Plan Limitations on Y2K Coverage," *ComputerWorld*, September 2, 1997 (quoting Nancy P. James, owner, N. P. James Agency in Concord, MA).

6. Eugene F. Wolters, *The FC&S Answer, Property & Casualty/Risk & Benefits Management*, April 27, 1998.

7. Robert Montgomery, Kelly A. M. Bowdren, Dwight Levick, Meike Olin, Karen P. Sartell, Kelly Cotter, Susanne Edes, Robert Dauer, and John C. Cross, "Managing Your Millennium Exposure," *The John Liner Letter*, vol. 35, no. 6, May 1998.

8. Caroline Saucer, A. M. Best editor, "Insurers Told to Rethink New Year 2000 Exclusions," *Best's Review*, May 18, 1998, 9 (L/H News, quoting Bill Malloy, managing director of J & H Marsh & McLennan, global broking).

9. Caroline Saucer, "Underwriting the Millennium," *Best's Review*, P/C, May 1997, 67 (interviewing Robert Omahne, executive vice president of New York–based American International Group, Risk Finance).

10. "Finding Risk Solutions in the Marketplace," Risk Strategist website, Risk-Focus Item, April 1997, www.jh.com/home/homepage.nsf.

11. Randy Myers, insurance business and finance writer, "Millenium [*sic*] Bug Coverage: Year 2000 Claims; Will Insurers Pay?" *Independent Insurance Agents Magazine*, December 1997, 35, 37, 39, 41.

12. Todd Muller, assistant vice president, Technical Affairs, Independent Insurance Agents of America, Inc., "ISO Y2K Filings," letter dated February 6, 1998.

13. Nancy P. James and Ann R. Truett, "Electronic Data: New Technology, New Risks," *John Liner Review*, Fall 1993.

14. "Insurance Issues for the Year 2000 Y2K Bug," Cooper, White & Cooper, LLP, Insurance Practice Group (San Francisco, CA) conference presentation volume, copyright 1999; conference held in San Francisco, November 17, 1999.

15. Michael Graham, partner at Barlow Lyde & Gilbert, "The Year 2000: How Non-Compliant Systems Affect Insurers," *Asian Insurance Review*, March 1998, 30, 32.

16. "Insurers Voice Concern Over Study's Estimate of Y2K Impact: AIA Says Speculation Could Have Adverse Impact; A Separate Study Suggests Costs Unpredictable," *The Standard*, July 9, 1999.

17. "Document Watch No.12," issued by An American Banker Newsletter Service, June 1998 (14 pages on insurance related to Year 2000 issues).

18. "Document Watch No.12."

19. "IIAA Softens Its Position on ISO Year 2000 Endorsements," *The Standard*, July 12, 1998.

20. "IIAA Softens Its Position on ISO Year 2000 Endorsements."

21. "Insureds, Brokers Want to Know How Cos. Will Treat Y2K Issues," *The Standard*, May 15, 1998, 1, 10–12. Philip Lian, senior vice president of Aon Risk Services and Eric Ball, senior vice president at Willis Corroon Americas, addressed the CPCU Society in a national satellite broadcast in May 1998.

22. "No Y2K Safety Net—Most Insurance Policies Have Holes," NewsEdge Corporation, August 11, 1998.

23. "Mass Division Raises Concerns Regarding ISO Y2K Exclusions," *The Standard*, July 10, 1998, 38.

24. J. M. Lawrence, "Linda Ruthardt: Served State as Insurance Commissioner," *Boston Globe*, March 5, 2010.

25. "Y2K Effects on Insurance Still Unknown," citing the Willis Corroon (risk management firm in Nashville, TN) 1999 insurance market forecast and quoting Kathleen Jensen (counselor on Y2K issues at the Association of Independent Insurers), via Penny Williams, *Insurance Times*, ca. March 24, 1999.

26. Scott M. Seaman and Eileen King Bower, "The Year 2000 Problem: The Good, the Bad, & the Ugly," *Mealey's Litigation Report: Insurance*, vol. 13, no. 33, July 7, 1999, footnote 1. In the article, it was noted that the Federal Information Processing Standard (FIPS) recognized the problem and "suggested" that agencies require a four-digit year field for software that was needed to deal with different centuries.

27. Patricia Vowinkel, "Insurers Seen Hit by Millennium Bug Losses," ZDNet, Ziff-Davis, Inc., June 4, 1998. Vowinkel quoted Jay Cohen, analyst at Merrill Lynch: "[I]f just 10 percent of the $600 billion fell into the laps of insurers, it would eat up about 20 percent of the U.S. insurance industry's $310 billion in policyholder surplus."

28. "Users, Insurers Grapple over Y2K," *InfoWorld*, August 18, 1998, via NewsEdge.

29. Myers, "Millenium [*sic*] Bug Coverage: Year 2000 Claims," 41.

30. Letter to all St. Paul agents from Gary P. Hanson, senior vice president of Sales and Marketing, The St. Paul, noting the creation of "information brochures to help you and your clients prepare for year 2000," accompanying the brochure "Year 2000: Is Your Future at Stake?" December 1, 1997.

31. USF&G Corporation, "Computers and the Year 2000," unsigned, with contact information for Bob Driscoll, USF&G Business Relationship Unit, December 1996.

32. Chris Mahoney, "Jan. 1, 2000 Already Bugging Risk Managers," *Insurance Times*, Special to the *Insurance Times*, April 1, 1997, 1, 15.

33. National Association of Professional Insurance Agents, *Y2K Year 2000 Survival Manual: Your Comprehensive Guide to Successful Technology Conversion and Management for Year 2000* (Alexandria, VA: PIA, 1998), sent by Patricia A. Borowski, division vice president, Research/Technical Affairs, PIA.

34. Sullivan Insurance Group, one-page untitled letter dated July 1999, signed Sullivan Insurance Group, Inc., received by author on July 14, 1999.

35. Martha R. Bagley, Bagley & Bagley PC, "Ready or Not . . ." *Women's Business*, August 1999, 6–7.

36. Helen Graves, "Going with the Y2K Flow," *Women's Business*, August 1999, 3.

37. Paul M. Yost and Paul E. B. Glad, Sonnenschein Nath & Rosenthal (San Francisco, CA), "Computing Coverage: Insurance Issues Arising Out of the Year 2000 Problem," *Insurance Litigation Reporter*, March 15, 1998.

38. "No Y2K Safety Net—Most Insurance Policies Have Holes" (quoting David Schaefer of Armfield, Harrison and Thomas, a Washington based insurance broker), NewsEdge Corporation, August 11, 1998.

39. Reuters New York via NewsEdge Corporation, January 15, 1999.

40. Robert P. Hartwig, chief economist, Insurance Information Institute, *National Underwriter*, August 30, 1999.

41. Sherin and Lodgen, LLP (10 Summer Street, Boston, MA 02110), "What Every Business Should Be Doing Now to Address the Y2K Problem," *Business Law Newsletter*, October 1998.

42. Rauer L. Meyer and Steven L. Hock, Thelen, Marrin, Johnson & Briggs, LLP, "Legal Issues and Risk of the Year 2000 Problem," The Year 2000 Information Center-Year 2000, June 20, 1998.

43. "Qualified Protection of Your Past Year 2000 Statements Requires Immediate Action by December 3, 1998," White and Williams, LLP (Philadelphia), to The St. Paul Companies, Re: Year 2000 Readiness Disclosures, signed by Thomas J. Ziomek, November 12, 1998.

44. "Countdown 2000: You, Your Independent Insurance Agent and Y2K Compliance," Independent Insurance Agents of America, Inc. (two white papers and short video), February 3, 1999.

45. "Countdown 2000: You, Your Independent Insurance Agent and Y2K Compliance."

46. "Countdown 2000: You, Your Independent Insurance Agent and Y2K Compliance."

47. "Year 2000: The Royal & SunAlliance Response," Royal & SunAlliance letter to the N. P. James Insurance Agency, signed by Paul H. Stewman, chief operating officer, Commercial Lines, July 15, 1998.

48. National Association of Professional Insurance Agents, *Year 2000 Survival Manual*, chapter 4: "Insurance Coverage and Client Issues," 37–43.

49. William C. Smith, "Big Prep for Y2K: Who Should Pay?: Businesses Spent Billions to Exterminate Bug, Want Insurers to Pony Up" (quoting Pittsburgh lawyer Terry Budd, counsel to Xerox in Y2K litigation against American Guarantee & Liability Insurance Co. for $183 million in remediation costs), *ABA Journal*, March, 2000, 88.

50. "Y2K: Avoiding a Wild West Scenario," *Insurance News Network*, February 9, 1999, via Penny Williams of the *Insurance Times*.

51. "Y2K: Avoiding a Wild West Scenario."

52. "To Our Producers," letter from Patrick G. Finckler, vice president, Marketing & Underwriting, Arbella Mutual Insurance Company (Quincy, MA), August 16, 1999.

53. Nancy P. James, N. P. James Insurance Agency (Concord, MA), "Modest Year 2000 Proposal: A Millennium Mediation Council: A Proposal to Avoid the Costs of Litigation Surrounding the Year 2000 Problem," guest editorial, *Insurance Times*, September 15, 1998, 7.

54. Sougata Mukherjee, *Business Journal*, Washington Bureau, December 18–24, 1998.

55. "Y2K Update: Preparing to Resolve Y2K Disputes," Robinson & Cole, LLP, April 1999.

56. James, "Modest Year 2000 Proposal: A Millennium Mediation Council."

57. "Y2K and Your Insurance: The Issues," J&H Marsh & McLennan, mid-1998, 35.

58. Saucer, "Underwriting the Millennium," 67.

59. "European Insurer Bucks U.S. Fears on Y2K," *Insurance Accountant*, June 8, 1998.

60. "Y2K: You Can't Insurance Your Way Out of This One," Microsoft Internet Explorer, June 20, 1998.

61. MSNBC staff and wire reporters, "Insurers, lawyers take Y2K action: 'Rude awakening' for business—and potential consumers," *MSNBC News* online, August 20, 1998.

62. Myers, "Millenium [*sic*] Bug Coverage." The author surveyed eight of the largest insurers: Cigna, CNA, Chubb, The Hartford, Ohio Casualty, The St. Paul, Travelers, and USF&G. All replied that they were undecided on either excluding or offering Y2K coverage.

63. "Y2K: Avoid Getting Hit by the Millennium Bug," brochure mailer, Royal & SunAlliance, May 1998.

64. "Insurance Issues Related to Year 2000 Exposures: Key Issues to Consider as We Approach the Data System Challenges Posed by the New Millennium," Aon Risk Services of the Americas, document version date: August 1998.

65. Myers, "Millenium [*sic*] Bug Coverage."

66. Myers, "Millenium [*sic*] Bug Coverage."

67. Stephanie Neil, "Y2K: You Can't Insure Your Way Out of This One," *PC Week*, March 6, 1998.

68. Thomas M. Reiter, partner in the law firm Kirkpatrick & Lockhart, LLP, "Policyholder's Guide to Coverage for Year 2000 Losses," *Journal of Insurance Coverage*, April 28, 1998, 51 (heavily footnoted, citing Campbell and Covington, and Pitts).

69. Joe Dwyer III, "Insurance Policies Provide Protection from Y2K Snafus," *St. Louis Business Journal*, September 7, 1998.

70. "Subject: Business Interruption Insurance," *Businesswire.com*, June 5, 1997 (From: Y2K Maillist To: year2000-discuss, forwarded to Nancy James by Michael Graesser, software technology consultant).

71. Myers, "Millenium [*sic*] Bug Coverage."

72. Alison Rea, "Does Your Computer Need Millennium Coverage?" *Business-Week*, March 10, 1997.

73. Reiter, "Policyholder's Guide to Coverage for Year 2000 Losses," 51.

74. Thomas Hoffman, "CIO's Wary of Year 2000 Insurance," *ComputerWorld*, December 1997, 1, 103.

75. Rea, "Does Your Computer Need Millennium Coverage?"

76. "Insurers Positioning Themselves to Avoid Damage from Computer Claims . . . 1997 . . . 1998 . . . 1999 . . . 2000 . . . 2001 . . . 2002," *Hartford Courant*, Business Weekly, September 22, 1997, quoting Lou Marcoccio, research director for the Year 2000 problems, the Gartner Group.

77. Kiplinger Washington Editors, "Dear Client" letter including Year 2000 commentary among other business-related items, *Kiplinger Washington Letter*, January 22, 1999.

78. "No Y2K Safety Net—Most Insurance Policies Have Holes," NewsEdge Corporation, August 11, 1998, citing a letter from William Kelly, senior vice president at J. P. Morgan and president of the International Federation of Risk and Insurance Management Associations (IFRIMA).

79. "ABA-Sponsored Insurance Program Offers Free Increased Cash Coverage for Y2K Exposure," *NEWS* (Washington), American Bankers Association press release, June 21, 1999, via fax to *Insurance Times*, forwarded to N. P. James by Penny Williams.

CHAPTER 5 CLAIMS AND LITIGATION

1. Edmund X. deJesus, BYTE senior technical editor, "Year 2000 Survival Guide," *BYTE* magazine cover story, July 1998, 52–62.

2. Laurence H. Reece, III and Joseph J. Laferrera, "Early Y2K Litigation in Mass. and Elsewhere," *Massachusetts Lawyers Weekly*, November 2, 1998, 11, 27.

3. C. Ramsay Raymond, owner, The Dreamwheel (Concord, MA), letter of July 13, 1999, to Nancy P. James, Owner, N. P. James Insurance Agency (Concord, MA).

4. Raymond, letter to James, July 13, 1999.

5. Terry Budd, and Curtis B. Krasik, Kirkpatrick & Lockhart, LLP (Pittsburgh, PA), "The Y2K Time Bomb—Policyholders Run for Coverage," American Bar Association, Committee on Insurance Coverage Litigation, coverage publication, March/April 1998, 3–17.

6. David M. Brenner, Seattle office of Graham & James, LLP, "Year 2000: Is a Coverage Meltdown Inevitable?" America Bar Association, Committee on Insurance Coverage Litigation, coverage publication, March/April, 1998, 48.

7. Sougata Mukherjee, "Year 2000 Problem Spells Trouble and Good Fortune," *Boston Business Journal*, November 14–20, 1997, 23.

8. "Solve Your Year 2000 Problem Now or Risk Being Sued," Year2000 Information Network online report, December 6, 1996, http://web.idirect.com/%7Embsprog/y2ksue.html.

9. Mukherjee, "Year 2000 Problem Spells Trouble and Good Fortune."

10. Daniel J. Goldstein, "Lloyd's Faces Millions in Y2K Claims," quoting Martin Leach, spokesman for Lloyd's of London and Matthew Jacobs, of Kirkpatrick & Lockhart, *Insurance Accountant*, March 16, 1998, 1, 4.

11. Scott Kirsner, Boston-based writer and consultant, often writing on Year 2000 issues, "Year 2000 Challenge: Keeping the Lawsuits at Bay," *CIO Magazine*, May 1, 1998.

12. Goldstein, "Lloyd's Faces Millions in Y2K Claims."

13. Kirsner, "Year 2000 Challenge: Keeping the Lawsuits at Bay," quoting attorney Doug Ey of Smith, Helms, Mulliss & Moore, LLP (Charlotte, SC).

14. Kirsner, "Year 2000 Challenge: Keeping the Lawsuits at Bay."

15. Jeff Jinnett, counsel for LeBoeuf, Lamb, Greene & MacRae (New York City), "Year 2000: Legal Issues Concerning the Year 2000 'Millennium Bug,'" 1996.

16. Rich Huggins, Rich Huggins and Associates (Palo Alto, CA), "The Year 2000 Hazard," *Tech Talk* (a publication of the Massachusetts Association of Insurance Agents), reprinted with permission of *Natural Hazards Observer*, vol. 23, no. 2, November 1998.

17. Alfred I. Jaffe, "Inevitable Year 2000 Lawsuits Put Insurance Carriers on the Hot Seat: Points & Viewpoints," *National Underwriter*, June 1, 1998, 47, 52.

18. "Round Table Group," *Business Wire*, March 2, 1999, via NewsEdge Corporation.

19. Lynda Radosevich, "Y2K Legal Games Begin," *InfoWorldElectric*, May 11, 1998.

20. Radosevich, "Y2K Legal Games Begin," referencing the Software Productivity Group Year 2000 conference held in New York in March 1998.

21. Joseph G. Blute, Mintz, Levin, Cohn, Ferris, Glovsky & Popeo (Boston), "R U OK if Y2K Claims Arise?" *Mass High Tech*, September 8–14, 1998. The article was adapted from Blute's recent Year 2000 seminar.

22. "Supreme Court Expert Testimony Case Could Affect Year 2000 Claims," *Insurance Times*, Federal Report, September 15, 1998, 4.

23. John Kerr, "Only a Test?" editor's letter, *Year 2000 Survival Guide Software Magazine* ("This Is a Test" issue), October 1998.

24. "Legislative Options on Y2K Suits Emerging," NewsEdge Corporation, May 12, 1998, via Penny Williams of the *Insurance Times*.

25. MSNBC staff and wire reporters, "Insurers, Lawyers Take Y2K Action: 'Rude Awakening' for Business—and Potential Consumers," *MSNBC News* online, August 20, 1998.

26. Ralph Ranalli, "Y2K Bill Verdict Eyed: Lawyers Hope for 'Bonanza' Clinton Veto Could Bring," *Boston Herald*, June 28, 1999, 26.

27. Ranalli, "Y2K Bill Verdict Eyed."

28. Ranalli, "Y2K Bill Verdict Eyed."

29. Ranalli, "Y2K Bill Verdict Eyed."

30. Dylan Mulvin, "Distributing Liability: The Legal and Political Battles of Y2K," *IEEE Annals of the History of Computing*, February 2020. Mulvin is assistant professor in the Department of Media and Communications, London School of Economics and Political Science.

31. Alex Maurice, "RMs Praise Y2K Law Provision," *National Underwriter*, July 26, 1999, 2, 39, quoting Lance Ewing, director of insurance and loss prevention for GES Exposition Services in Las Vegas and chairman of the External Affairs Team for the Risk and Insurance Management Society of New York; and Stephen Brostoff, "Clinton Inks Y2K Bill, Quells Insurer Fears," *National Underwriter*, July 26, 1999, 1, 38.

32. *Mealey's Litigation Report*, April 1998.

33. Mark Mehler and John Moore, "Are You at Risk?" Ziff-Davis, Inc., 1997–1998.

34. Kiplinger Washington editors, "Dear Client" letter, *Kiplinger Washington Letter*, January 22, 1999, which included Year 2000 commentary among other business-related items.

35. Paula M. Yost and Paul E. B. Gad, Sonnenschein Nath & Rosenthal (San Francisco), "Computing Coverage: Insurance Issues Arising Out of the Year 2000 Problem," *Insurance Litigation Reporter*, March 15, 1998, 164–174.

36. Reuters New York via NewsEdge Corporation, January 15, 1999.

37. Christopher Simon, Special to the *Wall Street Journal*, November 6, 1997, B12.

38. Simon, *Wall Street Journal*.

39. Maria E. Recalde, partner at Burns & Levinson LL (Boston), and chair of the Y2K Group, "The Final Countdown: Is Your Business Ready for the Year 2000?" *Women's Business*, August 1999, 8.

40. Deborah Chiaravalloti, senior vice president of Public Image Corp. (Newburyport, MA), "Managing Public Doubt: You Have a Responsibility to Minimize Y2K Fallout," *Women's Business*, August 1999, 9.

41. Mark W. Freel, Edwards & Angell, LLP (Providence, RI), "Across Frontiers: The Year 2000: Computer Bug Looms as the Millennium Approaches" (from *Counsellor*, Fall 1998), *Defense Research Institute*, Winter 1999, 5–6.

42. Freel, "Across Frontiers: The Year 2000: Computer Bug Looms."

43. Michael R. Cashman, partner with Zelle & Larson, LLP (Minneapolis), and Jeff Jinnett, counsel for LeBoeuf, Lamb, Greene & MacRae (New York City), "Insureds, Brokers Want to Know How Cos. Will Treat Y2K Issues," included in a national satellite broadcast address to the CPCU Society, May 1998, reported in *The Standard*, May 15, 1998, 1, 10–12.

44. John Jennings, "Technology, Y2K Issues Aired at Meeting," reporting on "IIC Miami Rendezvous" sponsored by Washington-based International Insurance Council, *National Underwriter*, January 18, 1999, 2, 7, 35.

45. David Reich-Hale, "'Black Hat' Seen Hurting Carriers on Y2K," *National Underwriter*, November 15, 1999.

46. Ian B. Hayes and William M. Ulrich, "The Tireless Case of Testing," *Software Magazine*, October 1998 ("This Is A Test" issue).

47. Joan Hartnett-Barry, "CAT CODE Y2K, Computer Date Error Threatens to Derail Business, Cause Claims," *CPCU Journal*, January 1998, 36–38, 79, 80. "CAT" is an insurance shorthand term meaning "catastrophe," which is a specific designation for measuring and assigning claims costs.

48. Irene Morrill, CPCU, CIC, ARM, LIA, CPIW, director of technical affairs, "Year 2000 Endorsements . . . Approved in Massachusetts," *Tech Talk*, October 1998.

49. Huggins, "The Year 2000 Hazard."

50. Ken Carson, head of Y2K practice group, Sugarman, Rogers, Barshak & Cohen, PC, "Y2K Law Alert: Legal Rights May Disappear Before the Y2K Witching Hour," May 1999.

51. Gregory P. Cirillo, partner, Williams, Mullen, Christian & Dobbins, Attorneys at Law (Richmond, VA), "Sequel: Y2K Remediation Contracts: When Your Back Is Against the Wall, You Do Not Need Leverage to Succeed," White paper, Sequel Group, August 20, 1998, Copyright 1998, Y2K, 1.1.c.

52. Cirillo, "Sequel: Y2K Remediation Contracts."

53. Barry D. Weiss and Ronald J. Palenski, Gordon & Glickson (Chicago), "Who Shall Be Answerable for Software Apocalypse?: Millennium-Related Computing Glitches Will Bring Judgment Day for Vendors and Users," *National Law Journal*, vol. 19, no. 11, November 11, 1996.

54. Trevor Thomas, "First Y2K Suits May Start Flood," *National Underwriter*, January 4, 1999, 3, 14.

55. Sender confidential, email, "Produce Palace Y2K Settlement," October 2, 1998.

56. "In a First, Parties Settle a Year 2000 Suit," *New York Times*, September 14, 1996.

57. "First Y2K Class Action Settlement Reported," *Mealey Publications* (King of Prussia, PA), October 13, 1998.

58. Thomas, "First Y2K Suits May Start Flood."

59. Thomas, "First Y2K Suits May Start Flood."

60. Kathleen Melymuka, "Here Come Year 2000 Liability Woes: Contractor Not Responsible for Missing Replacement Deadline," *ComputerWorld*, January 3, 1999.

61. Barnaby J. Feder, "Federal Court Is Asked to Rule on Year 2000 Insurance Dispute," *New York Times*, December 14, 1998, C5.

62. C. A. Soule, "Mass Y2K Suit Settled; Industry Watches Case Against Insurer," *Insurance Times*, December 1998.

63. Soule, "Mass Y2K Suit Settled; Industry Watches Case Against Insurer."

64. Feder, "Federal Court Is Asked to Rule on Year 2000 Insurance Dispute."

65. Arlene Polonsky, Morrison Mahoney & Miller (Boston), "Across Frontiers: Case Summary, IOWA: Insurance Company's Y2K Obligations," *Defense Research Institute*, Winter 1999, 8.

66. Barnaby J. Feder, "A Trickle of Year 2000 Lawsuits," *New York Times*, April 12, 1999, C4.

67. Bruce Telles, Rosenfeld, Meyer & Susman, LLP (Beverly Hills, CA), *Insurance Issues for the Year 2000*, Section 1: "Year 2000 Litigation: Flood, Trickle, or Mirage?" *Defense Practice Seminar Course Book* (Defense Research Institute, November 1999).

68. Thanks to Howard G. Zaharoff, Esq. and Northeastern law clerk, Jessica Andrade, Morse (CityPoint, 480 Totten Pond Road, 4th Floor, Waltham, MA 02451) for their research assistance in February 2024.

69. Chris Mahoney, "Mass. Suit Could Define Y2K Liability," *Insurance Times*, September 29, 1998, 1, 32–33.

70. Transcript of complaint: Michael J. Young, a representative of Anderson Consulting, LLP, a General Partnership, Plaintiff, and J. Baker, Inc., Defendant, August 28,1998, Norfolk, ss, Superior Court Division of the Trial Court Civil Action No. 98 01597.

71. BSC Consulting information sheet (Mitre Court, London EC1M 4EH), "Millennium Matters: News and Views on Year 2000 Embedded Systems Issues," June 1998.

72. David Tennant, partner at Nixon Peabody, LLP (Boston), "Y2K Coverage Wars Heat Up," *Mass High Tech*, December 20–26, 1999, 29.

73. Robert P. Hartwig, Insurance Information Institute's chief economist, *National Underwriter*, August 30, 1999.

74. Y2K Lawsuits Greet Insurers in New Year: Big Firms Suing Insurers to Recover Remediation Costs," *Insurance Times*, December 21, 1999.

75. "GTE Sues Insurers Over Y2K Repairs," *Newsbytes*, July 5, 1999, via NewsEdge Corporation.

76. "Y2K Lawsuits Greet Insurers in New Year: Big Firms Suing Insurers to Recover Remediation Costs."

77. Barnaby J. Feder, "GTE Sues 5 Insurers in a Bid to Spread Year 2000 Costs," *New York Times*, July 2, 1999.

78. Ara C. Trembly, "Barristers Battle Over Y2K Liability," *National Underwriter*, July 26, 1999, 2, 38.

79. Trembly, "Barristers Battle Over Y2K Liability."

80. Benjamin Love, principal and owner of Y2K Consulting and Insurance Mediation Practice (Dallas-Fort Worth), "Year 2000: A Road Map to CGL Coverage Issues." Article proof for *CGL Reporter*, Fall 1998, was sent to Barbara O'Donnell, Sherin and Lodgen, LLP (Boston), on November 3, 1998, for editing. Love previously served as inside counsel to insurers, TIG Insurance Company, United States Fidelity and Guaranty, and Fireman's Fund Insurance Company.

81. Yost and Gad, "Computing Coverage: Insurance Issues Arising Out of the Year 2000 Problem."

82. J&H Marsh & McLennan, *Y2K and Your Insurance: The Issues*, 1998, 7.

83. Walter J. Andrews and Leslie A. Platt, Wiley, Rein & Fielding (Washington, DC), "Insurance Coverage for the Millennium Bug," American Bar Association, Committee on Insurance Coverage Litigation, Coverage publication, March/April 1998, 47, note 4.

84. Bruce N. Telles, Rosenfeld, Meyer & Susman, LLP (Beverly Hills, CA), "The Danger of Asserting Fortuity as a Defense to Y2K Claims," *Defense Research Institute*, Covered Events, Insurance Law Committee Newsletter, Winter 1999, via Penny Williams, *Insurance Times*.

85. Robert Carter, McKenna & Cuneo, LLP (Washington, DC), on Y2K litigation, *ABA Journal*, September 2000.

86. E. E. Mazier, *National Underwriter*, January 8, 2001 (reported at $183 million remediation expenses); Mark Hollmer, "Y2K Coverage Questions Remain After Xerox Ruling," *Insurance Times*, January 9, 2001, 1, 12 (reported at $138 million remediation expenses); Thomas Hoffman, "Property Insurers May Face Y2K Suit Barrage," *Computerworld*, July 6, 1999, http://computerworld.com/cwi/story (reported at $183 million remediation expenses); Lee Copeland, "New York Judge Rules Against Xerox in Y2K Insurance Case," *Computerworld*, December 2, 2000, http://computerworld.com/cwi/story (reported at $183 million remediation expenses); John T. Hayes, Hayes Law Reports (Wilmington, DE, hayesreports.com), "N.Y. Court Rules Against Xerox on Y2K Insurance Coverage, Says Hayes Law Reports," courtesy of individual.com, December 21, 2000 (reported at $138 million remediation expenses); Barnaby J. Feder,, "New York Court Rules for Insurer on Year 2000 Claim," *New York Times*, December 22, 2000 (reported at $138 million remediation expenses).

87. Mazier, *National Underwriter*, January 8, 2001.

88. Robert P. Hartwig, chief economist, Insurance Information Institute, *National Underwriter*, August 30, 1999.

89. Peter Kelman, computer law specialist, Perkins, Smith & Cohen, LLP, "Companies Use an Old Maritime Clause to Sue Insurers," *High-Tech Quarterly, Insider View*, November 12–18, 1999.

90. Robert P. Hartwig, "Insurance Still Targeted for Y2K," *The Standard*, April 14, 2000.

91. Gavin Souter, "Lawyer Says Order May Aid Y2K Claims," *Business Insurance*, January 19, 1998, 3, 25.

92. Scott M. Seaman and Eileen King Bower, "The Year 2000 Problem: The Good, the Bad, and the Ugly," *Mealey's Litigation Report: Insurance,* vol. 13, no. 33, July 7, 1999, footnote 1.

93. Seaman and Bower, "The Year 2000 Problem: The Good, the Bad, and the Ugly."

94. Online article reprint,"Y2K to Be Top Issue in Directors and Officers Liability," *Business Insurance,* from *Best's News*, A. M. Best Company, December 7, 1998.

95. "Y2K to Be Top Issue in Directors and Officers Liability."

96. Warren S. Reid, WSR Consulting Group, specializing in management, technology, and litigation consultancy (Encino, CA), "2001: Legal Odyssey (The Millennium Bug)," *Computer Lawyer, Emerging Issues*, vol. 13, no. 6, June 1996, thanks to Barbara O'Donnell, Sherin and Lodgen, LLP (Boston).

97. Ara Trembly, *National Underwriter*, October 25, 1999.

98. David S. Godkin, Esq., "Government Agencies Offer Year 2000 Guidance, SEC and DOL Address Portfolio Company and Fund Compliance," *Venture Update*, Testa, Hurwitz & Thibeault, LLP, Summer 1999.

CHAPTER 6 LEGISLATION AND GOVERNMENT ON YEAR 2000 ISSUES

1. Penny Williams, "Millennium Liability Slated to Hit Insurers, Agents, Lawyers and More" (interview of Nancy James), *Insurance Times*, March 3, 1998, 1.

2. Robb Frederick, Ottaway News Service, "State, Federal Agencies in Race to Beat the Year 2000 Clock," *The Herald* (Sharon, PA), October 5, 1996, via Barron's.

3. Mark Crawford, "Year 2000 Software Fix Unlikely to Beat Clock," *New Technology Week* (Washington, DC), September 16, 1996.

4. "Agencies Urged to Prevent Computer Clock Crisis in 2000," Associated Press via Barron's, *Latrobe Bulletin* (Latrobe, PA), September 11, 1996; "Computer Crisis in 2000?" Associated Press via Barron's, *Minot Daily News* (Minot, MD), September 11, 1996.

5. Robert Sam Anson, "12.31.99, The Y2K Nightmare," *Vanity Fair*, January 1999.

6. Jube Shiver Jr., "Clinton Says Year 2000 Bug Is Solved for Social Security," *Boston Globe*, December 28, 1998, 1, 10. The SSA has a very large and active constituency in Americans over the age of sixty-five. Politicians at every level mess with that constituency at their own peril.

7. U.S. Senate, Special Committee on the Year 2000 Technology Problem, "Investigating the Impact of the Year 2000 Problem: Summary of the Committee's Work in the 105th Congress," February 24, 1999, Executive Summary, S. Prt. 106–10.

8. Jim, Abrams, "Senate Y2K Panel Reports That Seven States in 'Danger Zone,'" *Boston Globe*, September 23, 1999, A24.

9. "Opening Statement, Sen. Robert Bennett (R-Utah), Chairman, Special Committee on the Year 2000 Technology Problem Hearing on the Risks of Y2K on the Nation's Power Grid," press release, June 12, 1998.

10. "John Westergaard Testifies Before the U.S. Senate Select Committee on the Year 2000 Technology Problem," conducted at Foley Square Courthouse, New York City, *Year 2000 Washington*, July 6, 1998.

11. Cohen & Havian, LLP (Boston), "Tax Issues and the Year 2000 Problem," *CPA Client Bulletin*, January 1998, 3.

12. "Year 2000 Disclosure Bill Seen Reducing Liability for Firms," *Insurance Times*, Federal Report, September 15, 1998, 4.

13. U.S. Chamber of Commerce, Media Relations Department, "U.S. Chamber: Fix Y2K without the Lawsuits," *NEWS*, September 16, 1998, quoting Lawrence Kraus, president of the Chamber's Institute for Legal Reform.

14. "Year 2000 Disclosure Bill Seen Reducing Liability for Firms."

15. OpinionExchange, "Editorial Opinion: Time Out" (also referencing N. P. James's Millennium Mediation Council), *Insurance Times*, September 15, 1998, 6.

16. Jeri Clausing, "Year 2000 Compromise Bill Is Opposed by Trial Lawyers," *New York Times*, September 21, 1998.

17. Barnaby J. Feder, "Companies Still Hesitate to Share Year 2000 Information," *New York Times*, December 7, 1998.

18. Hill & Barlow, "Year 2000 Bulletin: Year 2000 Information and Readiness Disclosure Act," November 1998.

19. José A. Isasi, II, Rivken, Radler & Kremer (Chicago), "Across Frontiers: Millennium Bugged? A New Law Tells When a Disclosure Is Not a Disclosure," *Defense Research Institute*, reprinted from *Tort Source* (ABA), Winter 1999, 6–7.

20. "Washington DC, U.S.A." *Newsbytes* via NewsEdge Corporation, February 10, 1999.

21. *Newsbytes* reported that Don Meyer, spokesperson for the Senate Special Committee on the Year 2000 Technology Problem, stated that the proposed "Y2K Act" would not address lawsuits filed by companies against their insurers. "GTE Sues Insurers Over Y2K Repairs," *Newsbytes* via NewsEdge Corporation, July 5,

1999. The *New York Times* reported the same; see Barnaby J. Feder, "GTE Sues 5 Insurers in a Bid to Spread Year 2000 Costs," *New York Times*, July 2, 1999.

22. "Y2K Liability Limits," editorial, *New York Times*, July 3, 1999.

23. Edmund X. deJesus, *BYTE* senior technical editor, "Year 2000 Survival Guide," *BYTE* magazine, July 1998, 52–62.

24. Nancy P. James, N. P. James Insurance Agency (Concord, MA), "Modest Year 2000 Proposal: *A Millennium Mediation Council*: A Proposal to Avoid the Costs of Litigation Surrounding the Year 2000 Problem," *Insurance Times*, September 15, 1998, 7.

25. "Executive Order 13073—Year 2000 Conversion," Administration of William J. Clinton, February 4, 1998, 1–2.

26. "Year 2000: SEC Promises CEO Alert, Guidance Soon to Improve Inadequate Corporate Disclosure," *Federal News*, vol. 30, no. 28, July 10, 1998, 1046.

27. Marcy Gordon, Associated Press, "SEC Hands Down Y2K Ultimatum: Will Close Brokerages That Aren't Ready for Bug," *Boston Globe*, July 28, 1999; Robinson & Cole, LLP, "Y2K Update: Y2K Readiness—Focus on Securities Law," April 1999.

28. Barnaby J. Feder, "S.E.C. Guidelines to Yield Data on Year 2000 Risks," *New York Times*, October 5, 1998.

29. Alex Maurice, "Potential Flaws Cited in Y2K Liability Law," *National Underwriter*, August 30, 1999.

30. Maurice, "Potential Flaws Cited in Y2K Liability Law."

31. Maurice, "Potential Flaws Cited in Y2K Liability Law."

32. deJesus, "Year 2000 Survival Guide."

33. Quoted in Robert Pear, "Computer Trouble Looms for States in 2000, U.S. Finds: Social Welfare Programs Seen as Vulnerable to Delays If Repairs Are Not Made," *New York Times*, November 27, 1998, A1, A23.

34. "Washington DC, U.S.A.," *Newsbytes* via NewsEdge Corporation, February 10, 1999.

35. David E. Rosenbaum, "Vexing Party, Clinton Backs Year 2000 Bill: Law Would Put Limits on Punitive Damages," *New York Times*, June 29, 1999.

36. Anne E. Kornblut, "Senate Y2K Watchers Sound Muted Alarm," *Boston Globe*, March 3, 1999, A3.

37. "Washington DC, U.S.A."

38. Patrick Thibodeau, "Government Report Finds Few Y2K-Related Lawsuits," *ComputerWorld*, September 25, 2000.

39. "Year 2000 Computing Crisis: Strong Leadership and Effective Partnerships Needed to Mitigate Risks," statement of Joel C. Willemssen, Director, Civil Agencies Information Systems Accounting and Information Management Division, GAO/T-AIMD-98-276, United States General Accounting Office (GAO), testimony before the Subcommittee on Government Management, Information and Technology, Committee on Government Reform and Oversight, House of Representatives, September 1, 1998.

40. "Agencies Urged to Prevent Computer Clock Crisis in 2000."

41. Ellen Perlman, "A Lot of Government Computers Are Going to Crash as the Millennium Begins. Will Their Governments Crash Along with Them?" *Governing*, September 1996.

42. Perlman, "A Lot of Government Computers Are Going to Crash as the Millennium Begins."

43. "Y2K on the Net," *Governing*, September 1996.

44. Steve LeBlanc, CNC Statehouse Bureau, "Fargo Forum Urges Y2K Planning," *Concord Journal*, February 11, 1999, 11.

45. *Boston Globe*, December 30, 1998.

46. Joseph J. Chessey Jr., deputy commissioner, Massachusetts Department of Revenue Division of Local Services, "Year 2000 Compliance and Municipal Liability," *Bulletin to Local Officials*, October 1998.

47. Carolyn E. Boviard, director, Commonwealth of Massachusetts Department of Economic Development, "Conversion 2000: Y2K" kit, November 2, 1999.

48. Christa Degnan, "Beacon Hill Bitten by Year 2000 Bug," lead article, *Mass High Tech,* March 23–29, 1998. Note: Republican Charlie Baker was later to become governor of Massachusetts and the most popular governor in the country twenty years later.

49. *The Standard*, December 12, 1997.

50. "To: All Licensed Property & Casualty Insurance Companies," from Richard D. Rogers, deputy director, Consumer Division, State of Illinois Department of Insurance, re: "Year 2000 Coverage and Endorsements," April 1, 1998, issued by *An American Banker* Newsletter Service, "Document Watch No. 2"; and *Insurance Times*, June 15, 1998, 13.

51. David Turner, "The Illinois Compromise, Y2K Solutions," *Independent Agent*, April 1998, 23–24.

52. Eric Luenig, "Bill Would Curb Year 2000 Suits," www.news.com, April 17, 1998.

53. David M. Nadler, Dickstein Shapiro Morin & Oshinsky, LLP (Washington, DC), "State Governments Move to Limit Year 2000 Liability," *Newsbytes* via NewsEdge Corporation, April 15, 1998.

54. Michael H. Adams, Florida correspondent, Tallahassee, "Florida Takes Heat Off Year 2000 Liability Suits," *National Underwriter*, June 21, 1999, 6.

55. "Year 2000 Coverage and Endorsements," outline of the Division of Insurance actions, *Massachusetts Agent,* October 8, 1998.

56. "NCOIL Endorses Year 2000 Moratorium for Insurers," Press Release, *NAII News*, July 14, 1998. NCOIL is the National Conference of Insurance Legislators.

CHAPTER 7 ISSUES, PROBLEMS, AND LITIGATION SOLUTIONS

1. Stephen M. Honig and Lawrence R. Kulig, Goldstein & Manello (Boston), "Year 2000 Compliance Letters," *Mass High Tech*, May 25–31, 1998, 27.

2. Thomas C. Palmer Jr., "Some Fear T Won't Be Y2K-Proof," *Boston Globe*, December 3, 1999.

3. Roger W. Sudbury, executive officer, MIT Lincoln Laboratory (researcher of solid-state devices for modern radars), in an interview with Nancy P. James, May 22, 1998.

4. Capers Jones, chairman, Software Productivity Research, Inc. (Burlington, MA), "Abstract: Probabilities of Year 2000 Damages," February 27, 1998.

5. Capers Jones, chief scientist, Artemis Management Systems, Inc., Software Productivity Research, Inc., "Abstract: Executive Risks from the Year 2000 Software Date Problem," February 6, 1999, copyright 1998–1999.

6. Jones, "Abstract: Executive Risks from the Year 2000 Software Data Problem."

7. Capers Jones, "Abstract: The Aftermath of the Year 2000 Software Problem," September 16, 1998, copyright 1997–1999.

8. Jones, "Abstract: The Aftermath of the Year 2000 Software Problem."

9. Capers Jones, "Abstract: Probabilities of Year 2000 Damages—Version 3," February 6, 1999.

10. Capers Jones, "Abstract: Year 2000 Metrics—Version 2.1," March 19, 1999.

11. Capers Jones, "Abstract: Year 2000 Contingency Planning for Municipal Governments," Version 5, April 7, 1999, copyright 1998–1999.

12. Capers Jones, "Abstract: How Serious Is the Year 2000 Software Problem?" May 18, 1999, copyright 1997–1999.

13. Andrew Pollack, "Media & Technology: Chips Are Hidden in Washing Machines, Microwaves and Even Reservoirs," *New York Times*, January 4, 1999, C17.

14. *Boston Globe* via Associated Press, December 1998.

15. Fred Kaplan, "Military on Year 2000 Alert: 2000 a Computer Time Bomb for Military," *Boston Globe*, June 21, 1998.

16. Barnaby J. Feder, "Dispute Over New Wrinkle in Problem of Year 2000," *New York Times*, undated clipping.

17. Mark K. Anderson, "Year 2000 with 103 Days to Go, Many Are Not Ready," *Boston Globe*, September 20, 1999, C5.

18. Anderson, "Year 2000 with 103 Days to Go, Many Are Not Ready."

19. Barnaby J. Feder, "Fear of the Year 2000 Bug Is a Problem, Too: Year 2000 Bug Meets People Problem: Surprising Early Outbreak of Panic," *New York Times*, February 9, 1999, 1; Year 2000 checklist included.

20. Feder, "Fear of the Year 2000 Bug Is a Problem, Too."

21. Lynda Gorow, "Texas City Ready for Chaos, Just in Y2K Case," *Boston Sunday Globe*, September 18, 1999, 1, A12.

22. Donna Gold, Lisa Provost, and Matthew Taylor, "In New England, Some Stockpile, Others Yawn," *Boston Sunday Globe*, September 18, 1999, A12.

23. Feder, "Fear of the Year 2000 Bug Is a Problem, Too."

24. Nancy P. James, N. P. James Insurance Agency (Concord, MA), "Modest Year 2000 Proposal: *A Millennium Mediation Council*: A Proposal to Avoid the Costs of Litigation Surrounding the Year 2000 Problem," *Insurance Times*, September 15, 1998, 7.

25. Sandra A. Sellers, "Mediate Away the Millennium Bug," *Legal Times*, August 3, 1998.

26. Sellers, "Mediate Away the Millennium Bug."

27. Sellers, "Mediate Away the Millennium Bug."

28. "Y2K Negotiations Agreed On/ Fourteen Major Companies to Mediate Before Litigating," *Business Insurance* via NewsEdge Corporation, online article reprint, December 9, 1998.

29. Lisa A. Romeo, business development coordination–New England for the non-profit American Arbitration Association, "Mediation/Arbitration Keep Y2K Disputes Out of Court," *Women's Business*, August 1999, 10, 25.

30. Danny Ertel and Jeff Weiss, principals in CMI/Vantage Partners (Cambridge, MA), consulting on better managing critical corporate relationships, "Public Forum: Surviving Y2k—Together," *Boston Globe*, undated.

31. Sougata Mukherjee, "U.S. Chamber Wants New Court for Y2K Litigation," *Business Journal*, December 18–24, 1998.

32. James F. Henry, president, CPR Institute for Dispute Resolution, a non-profit alliance, "Y2K Problem: Talk, Don't Sue," *New York Times*, letter to the editor, December 28, 1998.

33. Marcia Stepanek, covers Y2K issues, "A Better Vaccine Against Y2K Lawsuit Fever," *BusinessWeek*, January 11, 1999, 48.

CHAPTER 8 INSURANCE AND Y2K BEYOND THE UNITED STATES

1. Mike Farish, "Millennium Doom," *Business*, undated.

2. Published seminar report by London solicitors Withers, "Insurer Liability Warning on Millennium," *Insurance Day*, December 11, 1996.

3. Chris Quick, "Millennium Clauses Under Scrutiny," *Insurance Day*, January 28, 1998.

4. Quick, "Millennium Clauses Under Scrutiny."

5. Robert Winnett, "Insurers Turn Their Back on 2000 Bug," *Sunday Times*, London, July 26, 1998, Money, 4.

6. Patricia Vowinkel, London, “Insurers Seen Hit by Millennium Bug Losses,” ZDNet, Ziff-Davis, Inc., June 4, 1998, quoting Jay Cohen, analyst at Merrill Lynch: “[I]f just 10 percent of the $600 billion fell into the laps of insurers, it would eat up about 20 percent of the U.S. insurance industry’s $310 billion in policyholder surplus.”

7. “Millennium Matters: News and Views on Year 2000 Embedded Systems Issues,” BSC Consulting (London), information sheet, June 1998.

8. “Insureds, Brokers Want to Know How Cos. Will Treat Y2K Issues: Exposure Identification, Remediation, Scope Discussed in CPCU Broadcast; Unique Nature of Event a Challenge,” *The Standard,* May 15, 1998, 1, 10–12.

9. Reuters, via NewsEdge Corporation, January 14, 1999.

10. “A. M. Best’s Y2K Exposure Grid,” *Best’s Review*, January 1999.

11. Senate Special Committee on the Year 2000 Technology Problem, “Investigating the Impact of the Year 2000 Problem,” February 24, 1999, 141.

12. Barnaby J. Feder, “Group Rethinks Rating on Year 2000 Readiness,” *New York Times*, January 22, 1999; “Group Won’t Publicize Year 2000 Ratings,” *New York Times*, February 1, 1999.

13. James P. Bond, “We’ll All Suffer If We Don’t Help Developing Countries Fix Their Y2K Problems,” *Boston Globe*, February 1, 1999.

14. International Association of Insurance Supervisors (IAIS), “For Immediate Release: International Financial Services Organizations Draw Attention to Year 2000 Issue,” November 14, 1997.

15. Juan Enriquez, researcher, Harvard University’s David Rockefeller Center, “Y2K Fears Could Hit Third World Hard,” *Boston Sunday Globe*, January 7, 1999.

16. Cynthia R. Koehler, Morrison Mahoney & Miller (Boston), “Across Frontiers: Going Global: So Who’s Ready for the Year 2000?” *Defense Research Institute*, Winter 1999.

17. “Year 2000: Financial Supervisors Urge Governments to Play Greater Role in Preparations,” *Federal News*, vol. 30, no. 28, July 10, 1998, 1048; joint statement by the Joint Year 2000 Council established in 1997, representing the Basel Committee on Banking Supervision, the Committee on Payment and Settlement Systems, the International Association of Insurance Supervisors, and the International Organization of Securities Commissions.

18. Eric Schmitt, “Congress Told of Progress on Solutions to Year 2000,” *New York Times*, March 3, 1999.

19. Anne E. Kornblut, “Senate Y2K Watchers Sound Muted Alarm,” *Boston Globe*, March 3, 1999, A3.

20. Fred Kaplan, “Military on Year 2000 Alert: 2000 a Computer Time Bomb for Military,” *Boston Globe*, June 21, 1998.

21. Ilana Gerard, specialist in competitive intelligence, trend forecasting, and strategic planning, is founder and senior partner of the Gerard Group International, with associates in twenty countries. “Y2K: Realities and Contingencies: Prepare for the Strategic Consequences of Y2K,” *Women’s Business*, August 1999, 7.

22. Edith M. Lederer, "U.S.: World Better Prepared for Y2K," Associated Press, June 22, 1999, forwarded to N. P. James by Penny Williams, *Insurance Times*.

23. Matt Kelly, special sections editor, *Mass High Tech*, December 27, 1999–January 2, 2000.

24. "Millennium Briefs," *Boston Globe*, undated clipping, ca. mid-1999.

25. "Across Frontiers: International Developments, CHINA: Executive Responsibility and Y2K," *Defense Research Institute*, Winter 1999, 9.

26. Lederer, "U.S.: World Better Prepared for Y2K."

27. Senate Special Committee on the Year 2000 Technology Problem, "Gartner Group's Predictions of Failure for Countries," undated, 144.

CHAPTER 9 Y2K GAFFES, SCAMS, WIT, AND WISDOM

1. Michael Cohen, "The Year-End Is Near," *Boston Globe*, December 27, 1999, A15.

2. Tim Graham, "Doomsday Fears About Y2K," *TechWeek*, June 29, 1998.

3. "Vehicle Titles Give Mainers Y2K Surprise," *Boston Globe*, October 13, 1999.

4. Simson L. Garfinkel, "Hooked by the Y2K Bug: Is It a Real Problem, or Is This Just Another Case of Millennial Hysteria?" *Boston Globe*, October 8, 1998, referencing www.mille.org.

5. Garfinkel, "Hooked by the Y2K Bug."

6. Suzi Parker, contributor, edited by Alex Heard, *New York Times Magazine*, October 4, 1998, 29.

7. Kevin Poulson, "The Y2K Solution: 'Run for Your Life!!'" *Wired*, August 1998, 122–25, 164–67. "They were hand-picked to kill the Millennium Bug. They hunkered down and started cranking out code. Now they're scared shitless."

8. Charles Platt, "America Offline (Inside the Great Blackout of '00)," *Wired*, August 1998, 165.

9. M. A. Nelen, "The Y2K Problem: How to Survive Meltdown AND the Lawsuits," *Mass High Tech*, July 13–19, 1998, 9.

10. Robert Huebner, COO of ONTOS, Inc., developer of e-commerce and web-enabled business solutions, "Public Forum: Y2K's Lessons for e-Commerce" (undated, unsourced clipping, ca. early autumn 1999).

11. Steven R. Avellino, "As You Were Saying . . ." column, *Boston Herald*, undated, ca. July 1999.

12. Ross Kerber, "Key Sectors in N.E. Seen Y2K-Ready," *Boston Globe*, December 27, 1999, "Y2K Countdown."

13. "Expert Urges Limits on Y2K Liability," citing a paper published by the Washington-based Progress & Freedom Foundation, "Briefs," *National Underwriter*, undated, ca. mid-1999.

14. Lisa S. Howard, London editor, "Insurers Might Impose Y2K Exclusions to Ease Exposure," *National Underwriter*, January 11, 1999, 1, 20, interviewing Simon Bird, managing director, underwriting, for HIH Casualty and General Insurance, Ltd. (London).

15. Editorial, "Post-Millennial Lawsuit Fever Won't Grip Lawyers—Really," *Mass High Tech*, January, 25–31, 1999.

16. Adam L. Penenberg, "Insurers Offer Millennium Bug Protection," *Wired News*, June 20, 1998.

17. *Boston Globe*, December 30, 1998.

18. Quote by Julie Bane, Miller Shandwick Technologies (Boston), "Close-up Report: Year 2000 Reader Predictions," *Mass High Tech,* December 27, 1999–January 2, 2000, 11.

19. Quote by Barbara Hemingson (Waupaca, WI), "Close-up Report: Year 2000 Reader Predictions," *Mass High Tech,* December 27, 1999–January 2, 2000, 11.

20. Barnaby J. Feder, "Intel and Compaq Dispute One Year 2000 Bug Theory," *New York Times,* January 11, 1999, reporting on Jace Crouch, a history professor in Michigan, and Michael Echlin, a Canadian programmer.

21. Mark Haselkorn, principal investigator, Policy and Global Affairs; Computer Science and Telecommunications Board; Division on Engineering and Physical Sciences; National Research Council, *Strategic Management of Information and Communication Technology: The United States Air Force Experience with Y2K* (Washington, DC: National Academies Press, 2007), https://doi.org/10.17226/11999. Note: this was one of less than a handful of documents obtained in 2020 and after for this project. The research conclusions of this U.S. Air Force report point to the extraordinary success of the nation's Y2K efforts.

22. Robert Kuttner, coeditor of the *American Prospect*, "Villains in the Y2K Case," *Boston Globe* (undated clipping, ca. mid-1999).

23. Barnaby J. Feder, "For Year 2000 Gnats, A Scented 'Bug' Spray," *New York Times*, no date.

24. Ian Hayes, president of Clarity Consulting, Inc., Year 2000 and IT strategy consultant (South Hamilton, MA), "One Day in 2004," *Software Magazine* ("This Is A Test" issue), October, 1998.

25. Marilyn vos Savant, Guinness Book of World Records Hall of Fame for "Highest IQ," "Are You Worried About the Y2K Bug?" *Parade Magazine*, May 9, 1999, 18–19.

26. Ted Bunker, Business, Capital Focus, "Polaroid Makes Instant Enemy with Y2K Ad," *Boston Herald*, June 28, 1999, 25.

27. MSNBC staff and wire reporters, "Insurers, Lawyers Take Y2K Action: 'Rude Awakening' for Business—and Potential Consumers," *MSNBC News* online, August 20, 1998.

28. U.S. General Accounting Office, Washington, DC Information Management and Technology Division, B-247094, to The Honorable Howard Wolpe, Chairman, Subcommittee on Investigations and Oversight, House of Representatives,

February 4, 1992. Major contributors to this report: Michael Blair, assistant director, Sally Obenski, evaluator-in-charge, and Paula Bridickas, computer scientist.

29. Jon Huntress, consultant (Portland, OR), Year 2000 Difficulties Advisory Commission to the Senate Committee on Small Business; Chairperson for the Education and PR Committee, Portland Year 2000 Ready Users Group, "The Cassandra Project," July 8, 1998, guest article, http://millenia-bcs.com/gstauthor.htm.

30. Fred Kaplan, "Military on Year 2000 Alert: 2000 a Computer Time Bomb for Military," *Boston Globe*, June 21, 1998.

31. Jon Van, "Software Bugs Turning Deadly in Complex Era," *Chicago Tribune*, December 14, 1986, quoting Edward Yourdon, vice president of DeVry, Inc.

32. Kaplan, "Military on Year 2000 Alert."

33. "Users, Insurers Grapple Over Y2K," *InfoWorld* via NewsEdge Corporation, August 18, 1998.

34. "Users, Insurers Grapple Over Y2K."

35. "Users, Insurers Grapple Over Y2K."

36. Anick Jesdanum, "Y2K May Trigger Computer Viruses," Associated Press article reprinted in *Boston Globe*, undated, ca. late 1999.

37. Bob Drogin, "Test Shows US Vulnerable to Hackers," *Los Angeles Times* article reprinted in *Boston Globe*, undated, ca. late 1999.

38. "Quotables," *Software Magazine*, "This Is A Test" issue, October 1998.

39. Huntress, "The Cassandra Project."

40. Huntress, "The Cassandra Project."

41. Christa Degnan, "Hype and Trepidation Prevail at Y2K Gathering in Boston," *Mass High Tech*, November 24–30, 1997, 5, quoting Paul Mosher, product marketing manager of Mitel Corp. (Ottawa, Canada).

42. "Y2K Maillist," forwarded by Michael Graesser to Nancy James, March 14, 1997, with March 1997 emails of inquiry and responses about Year 2000 insurance. Needing to look into accounting arbitrage, I referenced *Other People's Money: The Real Business of Finance*, a book by John Kay (New York: PublicAffairs, 2015): "Accounting arbitrage yields profits at the expense of those who rely on the integrity of accounts. Enron was an extensive user of accounting arbitrage, and Arthur Anderson's involvement in auditing this process was the cause of the accounting firm's demise. J. P. Morgan and Citigroup each agreed settlements of around $2B to settle claims made by Enron investors who alleged they had been duped by misleading accounts facilitated by transactions the banks had arranged. . . . Arbitrage is therefore a significant contributor to the trading profits of financial institutions. . . . Shadow insurance refers to transactions ceding liabilities to captive reinsurers under less strict regulation to circumvent the regulatory capital requirement."

43. John Dodge, "Partial Solutions for Y2K Problems," *Boston Globe*, TechEdge, undated clipping.

44. Robert Sam Anson, "12.31.99 The Y2K Nightmare," *Vanity Fair*, January 1999.

45. Jon Nordheimer, "Ringing It in Roundly," *New York Times*, December 27, 1998, 16.

46. Ross Kerber, "Navy Withdraws Report on Y2K Sewer Failures," *Boston Globe,* December 21, 1999.

47. Joseph R. Perone, "Technology Company Selling Insurance for the Year 2000," New Jersey Online: Business News, July 29, 1998.

48. CPSR Y2K Working Group, Rumor Central, June 19, 1998, http://www.cpsr .org/program/y2k/rumors/sectors.html.

49. CPSR Y2K Working Group, Rumor Central, June 19, 1998, attributed to http://www.tax.org/fotw/f120297.htm (January 25, 1998).

50. CPSR Y2K Working Group, Rumor Central, June 19, 1998, attributed to http://www.yardini.com/cyber.html (undated).

51. CPSR Y2K Working Group, Rumor Central, June 19, 1998, attributed to Information Week, copyright 1997, 1998.

52. CPSR Y2K Working Group, Rumor Central, June 19, 1998, attributed to http://www.gcn.com, copyright 1997, 1998.

53. CPSR Y2K Working Group, Rumor Central, June 19, 1998, attributed to comp.risks Digest 18.96, http://catless.ncl.ac.uk/Risks.

CHAPTER 10 CONCORD, MASSACHUSETTS, AND YEAR 2000 READINESS

1. CCTV, Concord-Carlisle cable television, video of the January 22, 1999 Forum hosted by the Concord Chamber of Commerce and moderated by Nancy P. James.

2. Greg Turner, "Panel to Examine Year 2000 issue," *Concord Journal,* January 14, 1999.

3. "Let's Get Involved Curing the Y2K Bug," *Arlington Advocate*, editorial, February 18, 1999.

4. Gary North, "FEMA Holds Closed Conferences. Were You Invited?" *Gary North's Y2K Links and Forums—Mirror*, February 2, 1999, www.y2k-links.com/ garynorth/3903.htm, via Valerie Kinkade, Concord Neighborhood Network.

5. Copy of email to Valerie Kinkade, Concord Neighborhood Network, from Joe Lenox, Chief, Concord Fire Department, and Civil Defense head, January 29, 1999.

6. Letter from Nancy P. James to Richard T. D'Aquanni, Applied Resources Group, Inc. (Brookline, MA), August 12, 1999, regarding his utility advisory work and my disappointment that he would not be able to be a panelist on my September 23, 1999, Y2K Forum in Concord.

7. Steve LeBlanc, CNC Statehouse Bureau, "Fargo Forum Urges Y2K Planning," *Concord Journal*, February 11, 1999, 11.

8. Greg Turner, "Y2K Readiness Tested, Forum Participants Say Town Is Prepared," *Concord Journal*, September 30, 1999, 1, 15.

9. Turner, "Y2K Readiness Tested."

10. Turner, "Y2K Readiness Tested."

11. Turner, "Y2K Readiness Tested."

12. Capers Jones, Artemis Management Systems, Inc., Software Productivity Research, Inc., "Abstract: Year 2000 Contingency Planning for Municipal Governments," Version 5, April 7, 1999, copyright 1998–1999.

13. Jones, "Abstract: Year 2000 Contingency Planning for Municipal Governments."

14. Email to NEPOOL (New England Power Pool) Participant Committee, Y2K Oversight Committee, August 20, 1999, forwarded to Nancy James from Dan Sack, head of Concord Municipal Light Plant, on August 26, 1999; "Bently Nevada has a lock on turbine vibration monitoring systems (and thrust bearing wear detectors) in the industry," warning of a possible date rollback on both August 21,1999 and December 31, 1999.

15. Email, September 8, 1999, from Valerie Kinkade, regarding N. P. James's upcoming Forum and her inquiry from the Center for Y2K and the Society of Concord's Readiness.

16. Barnaby J. Feder, "Year 2000 Activists Share Tales of Public Apathy," *New York Times*, October 24, 1999, 18.

17. Leslie Anderson, "Communities Map Plans for Y2K 'What Ifs,'" *Boston Sunday Globe*, December 19, 1999, 1, 6.

18. Anderson, "Communities Map Plans for Y2K 'What Ifs.'"

19. Anderson, "Communities Map Plans for Y2K 'What Ifs.'"

CHAPTER 11 AFTER 01/01/2000: 20/20 HINDSIGHT

1. Peter J. Howe, "Millennium 2000," *Boston Sunday Globe*, January 2, 2000, A14.

2. Howe, "Millennium 2000."

3. Howe, "Millennium 2000."

4. Howe, "Millennium 2000."

5. Howe, "Millennium 2000."

6. "Satellite Post Fixed After Y2K Glitch," *Boston Globe*, January 13, 2000, from the Associated Press; John Diamond, "Y2K Bug Bit Pentagon Satellites at Key Time," *Boston Globe*, January 4, 2000.

7. Barnaby J. Feder, "Business Day," *New York Times*, January 4, 2000.

8. Fred Kaplan, "Military on Year 2000 Alert: 2000 a Computer Time Bomb for Military," *Boston Globe*, June 21, 1998.

9. Michael S. Hyatt, "The Reality of Y2K Failures," July 7, 1999, email document forwarded to David Sunfellow, Sunfellow to Ramsay Raymond, and Raymond to Nancy James, July 10, 1999. Hyatt is the author of *The Millennium Bug: How to Survive the Coming Chaos* (1998) and *The Y2K Personal Survival Guide* (1998).

10. Hyatt, "The Reality of Y2K Failures."

11. Hyatt, "The Reality of Y2K Failures."

12. Terence Hunt, *Boston Globe*, undated, ca. mid-January 2000, from the Associated Press.

13. "Industry Sails into New Millennium with No Major Problems," *Insurance Times*, January 18, 2000.

14. William C. Smith, "Big Prep for Y2K: Who Should Pay?: Businesses Spent Billions to Exterminate Bug, Want Insurers to Pony Up," *ABA Journal*, March 2000, 88.

15. Barnaby J. Feder, "Smooth Entry of 2000 Is a Puzzle: Experts Who Expected Computer Failures Are Left Wondering," *New York Times*, January 9, 2000.

16. Feder, "Smooth Entry of 2000 Is a Puzzle."

17. Barnaby J. Feder, "Year 2000 Rollover Problem, the Sequel," *New York Times*, February 28, 2000, C5.

ABOUT THE AUTHOR

Nancy P. James was educated as a theoretical mathematician, followed by her first professional decade as a computer systems analyst, ultimately consulting on pre-internet national networks. Concluding that technology comprehended the enormity of insurance systems just as inadequately as insurance comprehended the liability risks of technology, in 1982 she founded an insurance agency specializing in technology liability risks and risk management.

When the millennium and Y2K computer challenges gained traction, James, lead author of a pioneering early Y2K article for the *Boston Bar Journal*, was offered a place on the 1997 American Bar Association annual meeting's technology panel, and subsequently an invitation to address international insurance experts in London. National and local speaking engagements followed through the millennium turn. When January 1, 2000, arrived and passed apparently without issue, James's attention returned to internet risks and management. Crafting the nation's first internet liability insurance policy in 1997, her professional life continued until 2017 with the sale of her agency. Thousands of pre-millennium documents waited in three storage boxes for the writing of this book. For more than four decades, James has been widely published in the areas of computer systems, global technology, cyber risk management, and liability.

James shares life with her husband, Professor Rick Frese, in the Massachusetts towns of Concord and Rockport. She is an avid swimmer and tennis player, sings in her church choir, and serves in local efforts for equality and justice.

Learn more at www.nancypjames.com.